Chemistry

Food science

a Special study

14 082656 4

Nuffield Advanced Science

Project team

E. H. Coulson, formerly of County High School, Braintree (organizer)
A. W. B. Aylmer-Kelly, formerly of Royal Grammar School, Worcester
Dr E. Glynn, formerly of Croydon Technical College
H. R. Jones, formerly of Carlett Park College of Further Education
A. J. Malpas, formerly of Highgate School
Dr A. L. Mansell, formerly of Hatfield College of Technology
J. C. Mathews, King Edward VII School, Lytham
Dr G. Van Praagh, formerly of Christ's Hospital
J. G. Raitt, formerly of Department of Education, University of Cambridge
B. J. Stokes, King's College School, Wimbledon
R. Tremlett, College of St Mark and St John
M. D. W. Vokins, Clifton College

Editor of the *Special studies* J. G. Raitt

This book and its experimental work were developed by:

J. M. V. Blanshard, Food Science Laboratories, University of Nottingham
K. B. Carlisle, Loughborough College School
P. S. Coles, Education Department, Buckinghamshire
Miss Audrey Free, Nottingham High School for Girls
Dr J. A. Knight, Loughborough College of Education
Dr R. A. Lawrie, Food Science Laboratories, University of Nottingham
Dr H. G. Muller, Department of Food Science, University of Leeds
Dr B. R. W. Pinsent, Unilever Research Laboratory
J. G. Raitt, Nuffield Foundation Advanced Chemistry Project
D. K. Rowley, King Edward VII School, Lytham
Dr G. Stainsby, Department of Food Science, University of Leeds
Professor A. G. Ward, Department of Food Science, University of Leeds

A further contributor to the text was:

J. Lamb, Department of Food Science, University of Leeds

Chemistry

Food science

a Special study

Nuffield Advanced Science
Published for the Nuffield Foundation by Penguin Books

Penguin Books Ltd, Harmondsworth, Middlesex, England
Penguin Books Inc., 7110 Ambassador Road, Baltimore, Md 21207, U.S.A.
Penguin Books Ltd, Ringwood, Victoria, Australia

Filmset in 10 on 12 pt 'Monophoto' Times by
Keyspools Ltd, Golborne, Lancs
and made and printed in Great Britain by
C. Tinling and Co. Ltd, Prescot and London.

Design and art direction by Ivan and Robin Dodd
Illustrations designed and produced by Penguin Education

Contents

Foreword

Sixth form courses in Britain have received more than their fair share of blessing and cursing in the last twenty years: blessing, because their demands, their compass, and their teachers are often of a standard which in other countries would be found in the first year of a longer university course than ours: cursing, because this same fact sets a heavy cloud of university expectation on their horizon (with awkward results for those who finish their education at the age of 18) and limits severely the number of subjects that can be studied in the sixth form.

So advanced work, suitable for students between the ages of 16 and 18, is at the centre of discussions on the curriculum. It need not, of course, be in a 'sixth form' at all, but in an educational institution other than a school. In any case, the emphasis on the requirements of those who will not go to a university or other institution of higher education is increasing, and will probably continue to do so; and the need is for courses which are satisfying and intellectually exciting in themselves – not for courses which are simply passports to further study.

Advanced science courses are therefore both an interesting and a difficult venture. Yet fresh work on advanced science teaching was obviously needed if new approaches to the subject (with all the implications that these have for pupils' interest in learning science and adults' interest in teaching it) were not to fail in their effect. The Trustees of the Nuffield Foundation therefore agreed to support teams, on the same model as had been followed in their other science projects, to produce advanced courses in Physical Science, in Physics, in Chemistry, and in Biological Science. It was realized that the task would be an immense one, partly because of the universities' special interest in the approach and content of these courses, partly because the growing size of sixth forms underlined the point that advanced work was not *solely* a preparation for a degree course, and partly because the blending of Physics and Chemistry in a single advanced Physical Science course was bound to produce problems. Yet, in spite of these pressures, the emphasis here, as in the other Nuffield Science courses, is on learning rather than on being taught, on understanding rather than amassing information, on finding out rather than on being told: this emphasis is central to all worthwhile attempts at curriculum renewal.

If these advanced courses meet with the success and appreciation which I believe they deserve, then the credit will belong to a large number of people, in the teams and the consultative committees, in schools and universities, in authorities and councils, and associations and boards: once again it has been the

Foundation's privilege to provide a point at which the imaginative and helpful efforts of many could come together.

Brian Young
Director of the Nuffield Foundation

Introduction

This is one of five Special studies which form part of the Nuffield Advanced Chemistry course for the 16 to 18 years age range leading to the Advanced Level examination of the GCE. The Advanced Chemistry course is intended to be modern in content, experimental in basis, and to integrate as fully as possible the physical, inorganic, and organic aspects of the subject. It also attempts to illustrate principles by means of examples drawn from applied fields such as chemical industry, agriculture, and medicine.

The course consists of nineteen topics to be followed by all students, together with one Special study to be selected from five possible alternatives: *Biochemistry, Chemical engineering, Food science, Ion exchange,* and *Metallurgy.* The Special studies are intended to provide opportunities for using again the principles studied earlier in the course, for seeing them in new contexts, for appreciating their relevance to neighbouring subjects, and for learning something of the social and economic effects of chemistry. They also provide an extended range of experimental work.

The *Teachers' guide to the special studies* includes guides to all the Studies in one volume, to help the teacher in choosing which to undertake.

Chapter 1

Food and food science

What is food science?

In a primitive community the main activities of everyday life are concerned with acquiring, growing, preparing, cooking, and eating food. The shadow of starvation is ever present or only just round the corner. But today in the industrialized areas of the world the majority of the population has become divorced from the annual cycle of seed time and harvest, so that for many people food has become only a packet on the supermarket shelf or a meal served in a canteen or restaurant. Yet even in the wealthiest and most sophisticated western community every individual can only live and play his part if the right amounts of the right types of food equip his body for active living.

The sciences concerned with food cover every stage of food production and consumption from the beginnings, such as the genetics of agricultural seeds, to the ends, the response of the human body to the various nutrients. No one aspect can be considered properly in complete isolation from the others. The whole range can be divided into three. This is artificial but it enables each of the three sections to be studied coherently and brings out the inter-relation between them.

The first of these sections extends to the stage when the farmer offers his produce for use as food or as a food raw material. He has by then grown the cereals, fruit, and vegetables, reared the animals ready for slaughter, milked the cows, and collected the eggs. The applied sciences which aid the farmer up to this point are grouped together as the agricultural sciences, which include animal husbandry, soil science, agronomy, animal nutrition, agricultural botany, bacteriology, zoology, and veterinary science. The final section comes when the food is passed into the digestive tract. Once digestion has begun we enter the realm of the medical sciences: biochemistry, physiology, nutrition, and bacteriology.

The central region of study, after the raw food has been grown and before it begins to be digested, is the realm of food science, but it cannot be completely separated from agriculture and medicine. The main activities in food science are concerned with the chemistry, physics, and biology which is involved in the preparation of food for immediate or future use: its processing, preservation from decay and attack by organisms, and its storage. Finally, there are the problems of transporting the raw or prepared food in a nutritious and palatable state to where it is required.

Only in the last twenty years has food science become a science in its own right as the study of the whole field of the properties, preservation, and processing of raw foods, and of the behaviour of the finished food products, although for more than a hundred years there have been chemical, microbiological, physical, and other scientific investigations which would today be regarded as food science.

The bacteriology of canning, the chemical composition of foods, how to secure rapid heating and cooling of liquids such as milk in pasteurization, are a few examples of early investigations. Food science must continue to be based on the past and future discoveries of the sciences from which it has sprung. Like medicine and agriculture, it relies both on the physical sciences: physics, chemistry, mathematics, statistics, metallurgy, and polymer science, and on the biological sciences: botany, zoology, biochemistry, physiology, bacteriology, biophysics, and nutrition. It must be able to take what is needed from each of these branches of knowledge and then fuse them to give the complete picture of the changes occurring in foods from harvest to consumption.

Almost all food components, whether fats, proteins, carbohydrates, minerals, or vitamins, originate from living tissues. Most foods are chemically extremely complex, and also vary according to the species, breed or strain, age, and sex, of the organism involved. The character of foods and the biochemical changes which they undergo are steadily being unravelled by the application of the most modern analytical techniques, including chromatography, mass spectrometry, and electrophoresis.

Even after foods have been produced in forms which are suitable for human consumption, they are subject to numerous adverse chemical, physical, microbial, or parasitic factors which may cause spoilage of the product or diseases of the customer. The food scientist seeks methods of preventing them. This aim has been reinforced by the recent explosive development of labour-saving convenience foods which can greatly reduce the time spent in the kitchen.

To the consumer, the importance of food lies primarily in the extent to which appearance and colour, water-holding capacity and juiciness, texture and tenderness, odour and taste, can be appreciated in the few moments before or during chewing. Food scientists have shown what is meant, in precise scientific terms, by the attributes of eating quality, and they have indicated how these might be artificially improved.

Within the last three centuries, because of the striking developments made possible by industrial and scientific progress, mankind has increased his powers of producing food and of conquering disease, and the world population has grown tremendously. Consequently there must be greater and more efficient utilization of existing food sources and development of entirely new ones. These are further vital considerations with which food scientists will become increasingly involved.

In the future, relationships must be sought between appetite and the assimilation of food on the one hand, and normal and abnormal human development on the other. Artificial foods, which have been controlled with respect to the attributes of eating quality, nutrient content, and ease of assimilation, will be developed. Perhaps such developments may lead eventually to the elimination of senility in old age and to the attainment of a longer life span.

Food science, like the agricultural and medical sciences to which it is so closely related, cannot be totally divorced from the social sciences or from political considerations, since it is involved with matters of interest to each individual and to the whole community.

The chapters which follow should be read to grasp the overall picture of the subject. Although science progresses through detailed and exact investigations, the time and opportunity for this in food science will come later for those who may wish to study the subject further. For others the broad outline of the subject may help them to consider how an understanding of the chemistry, physics, and biology of foods can contribute towards the improvement of the health and well-being of mankind, and towards solving the problems of the shortage of food in the world.

What is food?

A food is a chemical substance which, when eaten, digested, and absorbed by the body, will promote the growth and repair of tissues, produce energy, and regulate these processes. The chemical components of food which perform these functions are nutrients. There are six major groups of nutrients: fats, proteins, carbohydrates, vitamins, mineral elements, and water.

A substance is only a food if it contains a nutrient. Some of the items which often form part of our diet are not actually foods. Pepper is widely used as a seasoning but it is in no way necessary to the human body, and it is therefore not a food. Tea and coffee are widely drunk and are valued by many people for their flavour and mild stimulant action which is due to the substance caffeine, but they are not foods. The value of these drinks lies in the water, and in the milk and sugar which may have been added to them. On the other hand, common salt *is* a food since sodium ions and chloride ions are essential to the body processes. Cocoa is a food since cocoa beans, and powder, contain fat, carbohydrate, and protein.

In addition to these nutrients, the body also requires a continuous supply of oxygen to release energy from the food.

The principal functions of nutrients are shown in table 1.1.

Growth and repair	**Provision of energy**	**Regulation of body processes**
proteins	fats	vitamins
mineral elements	carbohydrates	mineral elements
water	proteins	proteins (e.g. enzymes)
		water

Table 1.1
The principal functions of nutrients

Fats, carbohydrates, and proteins can all be used by the body to provide energy. The British Medical Association recommends that in the United Kingdom about 30 per cent of energy intake should be as fats and about 70 per cent as carbohydrates. In 1969 the Department of Health and Social Security published a list of recommended daily intakes of energy and nutrients, and some extracts from this are given in table 1.2. The figures in this table represent averages, and are useful in planning the food requirements of whole populations. People who lead apparently similar lives differ in their nutrient needs.

Age and sex	**Energy /MJ**	**Proteins /g**	**Iron /mg**	**Calcium /mg**	**Thiamine (a B vitamin) /mg**	**Nicotinic acid (a B vitamin) /mg equivalents**	**Ascorbic acid (vitamin C) /mg**
0–1 years							
infants	3.3	20	6.0	600	0.3	5.0	15
15–18 years							
boys	12.6	75	15.0	600	1.2	19	30
girls	9.6	58	15.0	600	0.9	16	30
Men, 18–35 yrs.							
moderately active	12.6	75	10.0	500	1.2	18	30
very active	15.1	90	10.0	500	1.4	18	30
Women, 18–55 yrs.							
most occupations	9.2	55	12.0	500	0.9	15	30
pregnancy, 2nd and 3rd trimester	10	60	15.0	1200	1.0	18	60
lactation	11.3	68	15.0	1200	1.1	21	60

Table 1.2
Recommended daily intakes of energy and some nutrients for the United Kingdom (NOT a complete list). *Department of Health and Social Security,* 1969

The list in table 1.2, though incomplete, raises the question of how we are to be provided with all the nutrients which are necessary for health. Some foods such as sugar provide only one nutrient. Other foods provide a wide range of nutrients; and amongst these the best is milk. A properly balanced diet should regularly contain a range of different foods, and in the United Kingdom these are likely to be drawn from five categories: (1) cereals (which includes bread); (2) milk, cheese, and eggs; (3) meat and fish; (4) fruit and vegetables; and (5) fats, sugar, and preserves. No group is adequate on its own, but by consuming some from each group we take in an adequate supply of the essentials.

We shall now examine in more detail the six classes of nutrients which a properly balanced diet will provide.

Fats and oils

Fats are esters of alcohols with aliphatic acids. The alcohol is usually glycerol, and the acids are usually palmitic, stearic, and oleic; but higher alcohols, and acids with several carbon to carbon double bonds, and phosphoric acid, also produce fatty substances. Since glycerol is a trihydric alcohol, several different esters may be formed with an acid RCO_2H. Three of these are shown below.

glycerol	monoglycerides	diglycerides	triglycerides
CH_2OH	CH_2O_2CR	CH_2O_2CR	CH_2O_2CR
|	|	|	|
$CHOH$	$CHOH$	CHO_2CR	CHO_2CR
|	|	|	|
CH_2OH	CH_2OH	CH_2OH	CH_2O_2CR

You may like to work out the formulae of the two other esters which exist in this series. They are isomers of compounds shown above.

Monoglycerides and diglycerides are formed during the course of the digestion of fats and in the synthesis of fats, but almost all the fats stored in plants or animals are triglycerides. The fatty acids are long straight chain molecules; the commonest have between 16 and 18 carbon atoms, and some of them are unsaturated. The three commonest fatty acids are shown below.

palmitic acid	$C_{15}H_{31}CO_2H$	
oleic acid	$C_{17}H_{33}CO_2H$	(unsaturated)
stearic acid	$C_{17}H_{35}CO_2H$	

The main fat in olive oil is the triglyceride of oleic acid, and is called triolein.

Most naturally occurring glycerides, however, are mixed glycerides, and contain two or more different acids in the molecule; this is particularly true of animal fats.

$$\begin{array}{ll} CH_2O_2CC_{17}H_{33} & CH_2O_2CR^1 \\ | & | \\ CHO_2CC_{17}H_{33} & CHO_2CR^2 \\ | & | \\ CH_2O_2CC_{17}H_{33} & CH_2O_2CR^3 \end{array}$$

triolein — a mixed triglyceride

Most natural fats are mixtures of mixed glycerides. The triglycerides are the main long term energy reserve of the individual cell, and especially of the animal body as a whole. Energy sufficient to maintain an animal for many weeks may be stored in this way. Triglycerides are poor conductors of heat, and another of their functions in the body is to provide excellent heat insulation for animals in cold climates.

Butter fat contains a large proportion of short chain fatty acids of which butyric acid, $CH_3CH_2CH_2CO_2H$, is the principal. These compounds contribute to the low melting point of butter.

The fats are one group of compounds in a larger group called the 'lipids'. All of the lipids are polar molecules, a feature which is significant in the structures which they form in living organisms.

The simpler fats largely constitute stored food, from which the organism may obtain energy. The more complex fats are found in structures such as cell membranes and nervous tissue such as the brain.

Proteins

Proteins are polymers whose monomer building units are the amino acids. Theoretically many thousands of amino acids could exist, and many have been synthesized, but it is a remarkable fact that only twenty are commonly found in nature. These are all α-amino acids, that is, the $—NH_2$ group is attached to the carbon atom adjacent to the $—CO_2H$ group.

$$\begin{array}{l} \quad\gamma \quad\;\; \beta \quad\; \alpha \\ RCH_2CH_2\underset{\displaystyle NH_2}{\underset{|}{C}}H—CO_2H \end{array}$$

an α-amino acid

The general formula for an α-amino acid is thus $RCH(NH_2)CO_2H$; and the simplest amino acid is when R is H—, giving glycine. When R is CH_3— the compound is alanine.

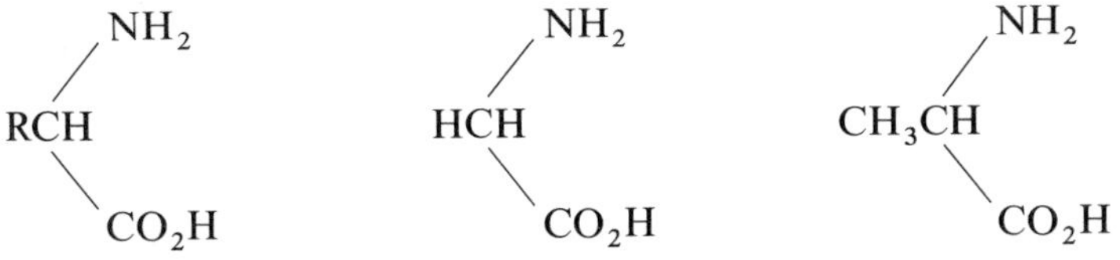

$$\underset{\text{general formula}}{RCH(NH_2)CO_2H} \qquad \underset{\text{glycine}}{HCH(NH_2)CO_2H} \qquad \underset{\text{alanine}}{CH_3CH(NH_2)CO_2H}$$

In alanine the α carbon atom has four different groups attached to it and it is therefore asymmetric. The compound is thus optically active and exists in D- and L-forms. With the exception of glycine all the twenty amino acids which normally occur in proteins are asymmetric and optically active, but it is a remarkable fact that they are almost always in the L-configuration.

Since amino acids contain both a basic and an acidic group they occur almost exclusively in an ionized form. If the acid is dissolved in water it will have the formula $^+NH_3RCHCO_2^-$. If it is in acid solution the —CO_2^- group will become —CO_2H, and if it is in alkaline solution the —$^+NH_3$ will become —NH_2. The situation can be summarized as shown below.

acid solution	distilled water	alkaline solution
$^+NH_3RCHCO_2H$	$^+NH_3RCHCO_2^-$	$NH_2RCHCO_2^-$

The net charges on the structures are +, 0, and − respectively. This is of great importance in the properties of the polymers of amino acids, the proteins. The pH at which the net charge is zero is known as the isoelectric point.

When the —CO_2H group of one amino acid condenses with the —NH_2 group of another amino acid, eliminating water, a dipeptide is formed.

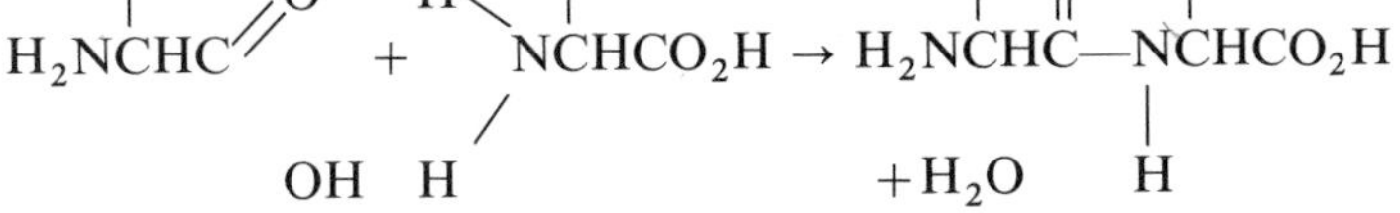

$$H_2NCH(R^1)C(=O)OH + H\text{—}N(H)CH(R^2)CO_2H \rightarrow H_2NCH(R^1)C(=O)\text{—}N(H)CH(R^2)CO_2H + H_2O$$

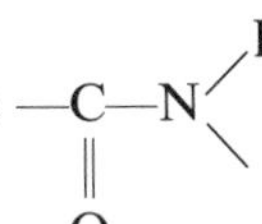

The linkage $\text{—}C(=O)\text{—}N(H)\text{—}$ by which the amino acids are joined is known as the peptide link. The molecule rotates until the C=O and N—H are *trans* to each other. It will be noticed that the dipeptide contains a $—NH_2$ group and a $—CO_2H$ group, and the process may therefore be repeated to build up a larger polymer. Those polymers which have molecular weights up to about 2000 are known as peptides. Above a molecular weight of about 5000 the naturally occurring polymers are known as proteins. The dividing line between peptides and proteins is diffuse.

The diagram below shows the sequence of amino acids in a protein, which is known as the primary structure.

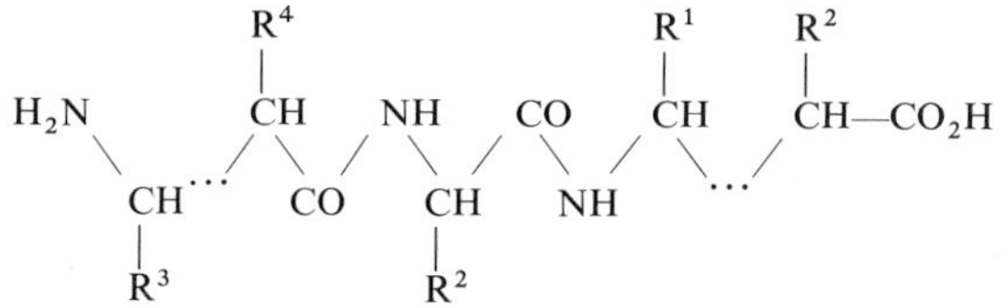

```
                R4                            R1      R2
                |                             |       |
H2N             CH     NH     CO              CH      CH—CO2H
   \           / \    /  \   /  \            / \     /
    CH ...    CO     CH     NH        ...
    |                |
    R3               R2
```

A representation of a polypeptide or a protein chain 'backbone'

If one of the R— groups contains a $—CO_2H$ or a $—NH_2$, then ionization can occur at this point. Since several of the naturally occurring amino acids do contain such groupings the phenomenon is widespread, and the ionization of proteins is primarily due to the ionization of the R— groups and not to the terminal $—CO_2H$ or $—NH_2$. Aspartic acid is an amino acid with a $—CO_2H$ in the R— group and lysine is one with a $—NH_2$ in the R— group.

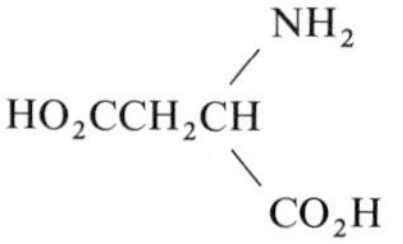

```
              NH2
             /
HO2CCH2CH                 aspartic acid
             \
              CO2H

               NH2
              /
H2N(CH2)4CH               lysine
              \
               CO2H
```

```
    R4                       R1
    |                        |
    CH     NH      CO        CH
   / \    /  \    /  \      /  \
...    CO      CH      NH        ...
               |
               CH2
               |
               CO2-
```

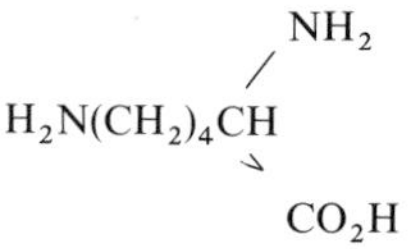

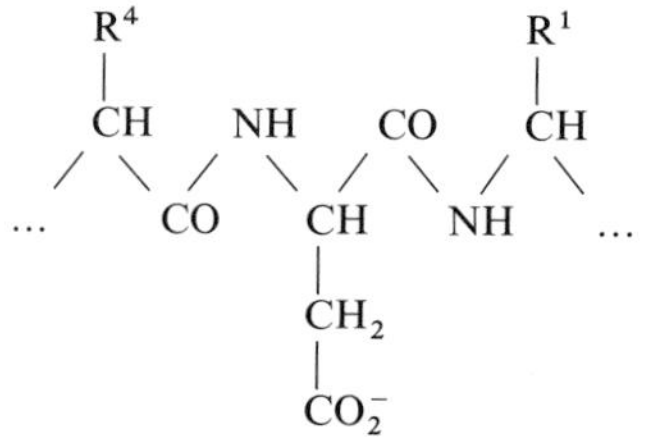

Part of a protein chain 'backbone' showing an ionized R— group (due to an aspartic acid residue)

The existence of charged side-groups gives rise to the possibility of frequent cross-linking between protein chains, and internal linking within one chain by means of electrostatic bonds. The number and nature of these side-groups determines the isoelectric point of a protein.

The polypeptide chain in a protein is not usually extended but is often coiled in a form resembling an extended spiral spring, known as the α-helix. The backbone is held in this form by means of extensive hydrogen bonding between the >N—H and O=C< groups in the chain, i.e. >N—H···O=C<. The helical formation is known as the secondary structure. The >N—H in one peptide link hydrogen bonds to the >C=O in the fourth peptide link away. In polyglycine the fit is perfect and all of the peptide groups can bond in this way; with other amino acids, and with mixtures, the fit is not perfect and this gives rise to deviations in shape from a perfect single spiral.

The thread formed by the helical structure may be stranded with other threads to form a rope, as in figure 1.1(*i*) or it may be folded and crumpled into itself in a whole variety of shapes as in (*ii*).

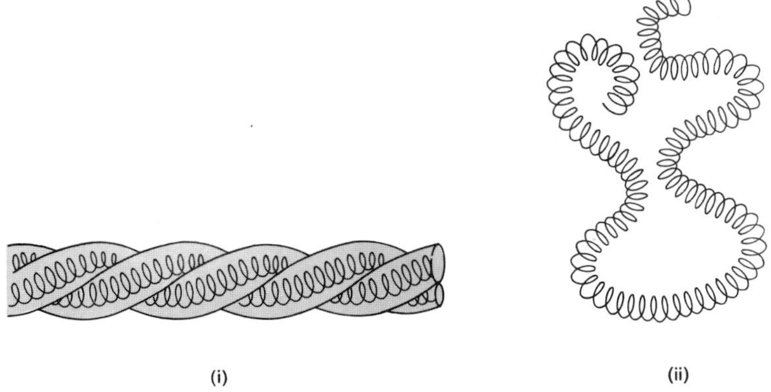

Figure 1.1
Two different types of tertiary structure in proteins: (*i*) fibrous proteins, (*ii*) globular proteins.

The configuration which the α-helix takes up in stranded form or in folded form is known as its tertiary structure. Stranded proteins form the fibres in the keratin of hair, collagen of tendon and bone, and myosin of muscle. The majority of proteins, however, are of the type shown in (*ii*), globular proteins; examples are egg albumin and blood serum albumin.

The secondary and tertiary structures of proteins are maintained principally by hydrogen bonds, and to some extent by ionic bonding between $—NH_3^+$ and $—CO_2^-$ groups. Hydrogen bonds are relatively weak and may be broken by heating, and the charges on $—NH_3^+$ and $—CO_2^-$ groups depend upon the pH of the medium. Heating (usually above 70 °C), and the addition of strong acids and alkalis, can therefore bring about the rupture of the links responsible for secondary and tertiary structure, and the structures disintegrate. This structural disintegration leads to a loss of the original chemical and physical properties, and the protein is said to have been denatured. Boiling or frying an egg leads to denaturation of the protein in the white of the egg, causing it to change from a transparent viscous liquid to a white opaque solid.

On denaturation, the primary protein structure remains unchanged; that is to say, the amino acid sequence in the backbone is unaltered. But the properties associated with the secondary and tertiary structure have gone, biological activity is destroyed, solubility decreases, and viscosity increases.

Proteins fulfil many of the most important functions in living cells. They can form part of body structure such as the collagen in bone or cartilage, act as a food store as does the albumin in egg, as buffer systems maintaining pH, and as oxygen carriers as in haemoglobin. But probably most important of all they perform catalytic functions in the form of enzymes.

All life processes are the sum of a vast complexity of chemical reactions, and the majority are catalysed by enzymes. Many of the syntheses achieved by enzymes have still not been copied by man. Moreover, enzymes have the remarkable property of being able to catalyse in aqueous solutions close to neutrality (within the range pH 5.0–8.0), and at very moderate temperatures, for instance 0 to 40 °C. If organic chemists had to work within such narrow limits, their achievements would be small indeed.

Some enzymes will catalyse several different reactions but most are very specific and will catalyse only one particular reaction. For instance, the enzyme which catalyses the oxidation of lactic acid to pyruvic acid will catalyse this reaction and no other; moreover, it will react only with the laevo-isomer of lactic acid and not with the dextro-isomer.

Only in a very small number of instances is much known about the mechanisms by which enzymes achieve their results. It is clear, however, that they combine temporarily with the substance or substances which they are going to modify, in the manner of jig-saw pieces fitting together, and they then separate into the changed product and unchanged enzyme. Thus the shape of the enzyme molecules is of central importance. Since the shape of a protein is destroyed on denaturing, enzymes are destroyed by excessive heat or strong acids and alkalis.

An adequate intake of proteins is required in order to maintain the growth and repair of tissues. Proteins are broken down by enzymes in the digestive tract into their constituent amino acids. These amino acids are then transported by the blood to the sites where they are required and there synthesized to the appropriate structure by other enzymes. The body proteins are themselves being continually broken down and resynthesized, and during life the proteins and amino acids are in dynamic equilibrium with each other. Half of the liver protein of a rat is resynthesized in a week. (This fact was determined by a radioactive tracer technique, using amino acids containing the radioactive ^{15}N atoms.)

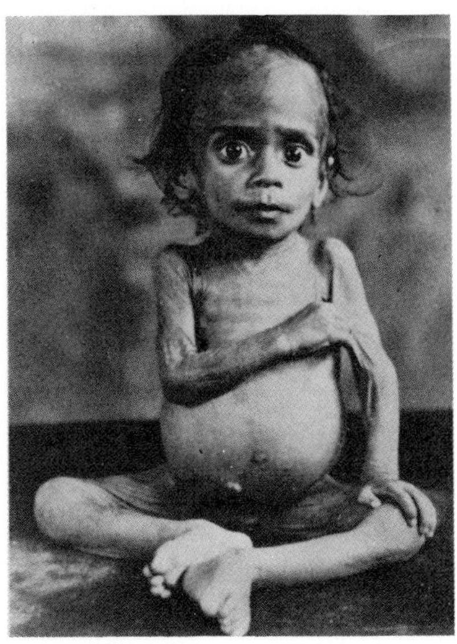

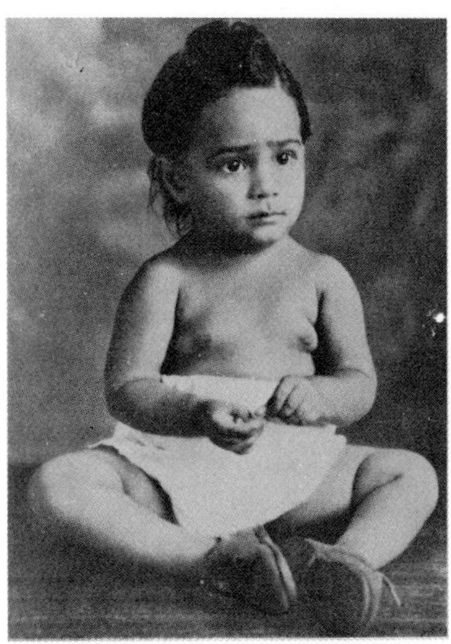

Figure 1.2
Nutritional Marasmus (combined protein-calorie deficiency) and recovery. (*i*) Child with Nutritional Marasmus showing extreme wasting of muscle and fat. (*ii*) The same child after nutritional recovery. Note the change in body proportions.
From Jelliffe, D. B. (1966), The assessment of the nutritional status of the community, *World Health Organization*, monograph 53.

Of the twenty or so amino acids which occur in foods only eight are essential to adult human beings, and two more appear to be necessary for healthy growth in infants. The implication of this is that the body cannot synthesize these ten compounds, but the remainder can be produced from the ten essential ones. Neither glycine nor alanine, which were mentioned earlier in this chapter, is essential. But the compound phenylalanine (closely related to alanine) is essential, and so is lysine.

$C_6H_5-CH_2CH(NH_2)CO_2H$ phenylalanine

$H_2N(CH_2)_4CH(NH_2)CO_2H$ lysine

Animal proteins usually provide the ten essential amino acids in proportions which are suitable for man. Plant proteins often lack one or more essential amino acids, though the soya bean has all the necessary amino acids. In general, the best use is made of protein foods if meat protein and plant protein are eaten in approximately equal amounts.

Carbohydrates

The simplest carbohydrates are called monosaccharides. Commonly known as sugars, they are crystalline substances which dissolve in water to give sweet solutions. An example is glucose, which exists as two isomers, α-D-glucose and β-D-glucose.

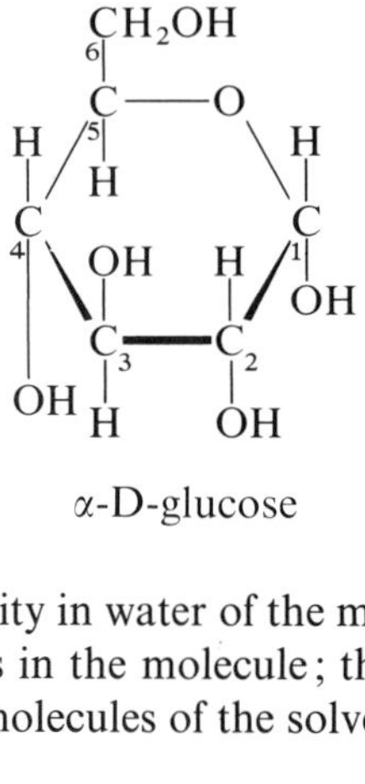

α-D-glucose 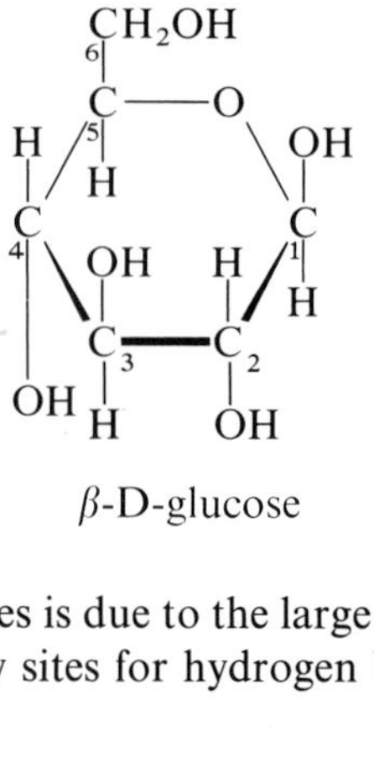 β-D-glucose

The high solubility in water of the monosaccharides is due to the large number of —OH groups in the molecule; this gives many sites for hydrogen bonding with the water molecules of the solvent.

From the diagrams above it will be seen that the structural difference between the two forms of glucose is in the arrangement of the —H and —OH groups on carbon atom number 1. Although this may appear a small difference it produces very important chemical and physical differences between the polymers made from these two monomers.

Other monosaccharides containing six carbon atoms are fructose and galactose.

Monosaccharides can combine to form disaccharides; the monomers unite by the elimination of a molecule of water from two —OH groups.

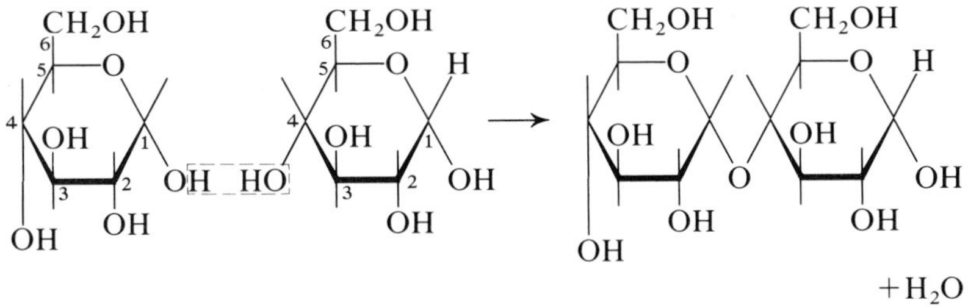

α-D-glucose α-D-glucose Maltose

$+H_2O$

The link is formed between carbon atom 1 on one molecule and carbon atom 4 on the other, and it is therefore known as a 1:4 linkage. Maltose is a disaccharide.

β-D-glucose dimerizes to give cellobiose.

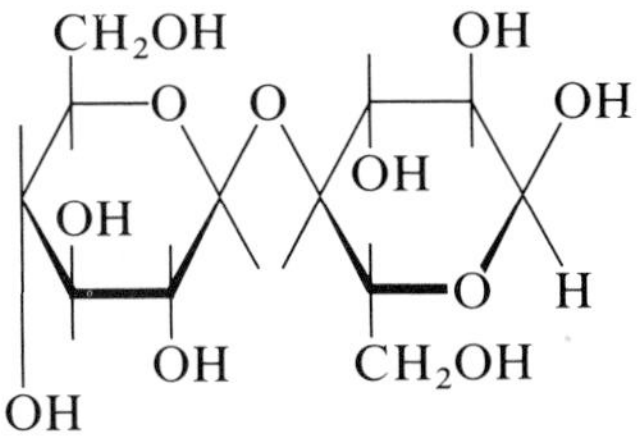

1: 4β link
cellobiose

Note that in this dimer the second glucose unit has been rotated through 180°.

Polysaccharides are polymers formed by linking very large numbers of monosaccharide molecules. Glucose is the most important monomer of the naturally occurring polysaccharides.

To form a polysaccharide we could continue to link α-D-glucose units by 1:4 linkages.

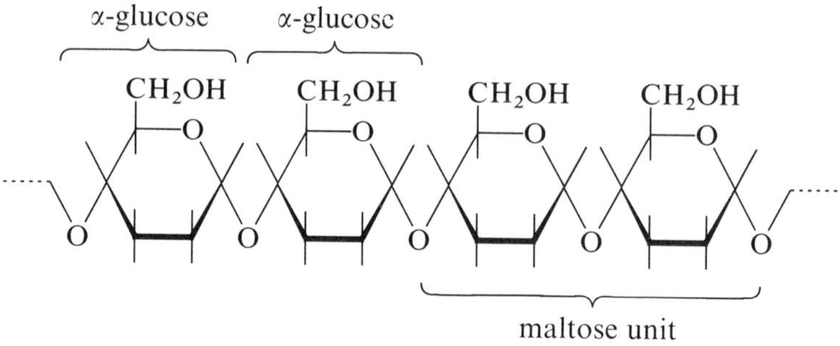

A simplified representation of α-D-glucose rings linked to form amylose

One of the components of starch, amylose, is formed by linking about 1000 to 4000 units of α-D-glucose to give a very long chain (with a molecular weight of about 160000 to 700000).

It is possible, however, for the 6-carbon atom of the α-D-glucose molecule to form a linkage with the 1-carbon of another glucose molecule. This gives a 1:6 linkage, which gives rise to the possibility of branched chains:

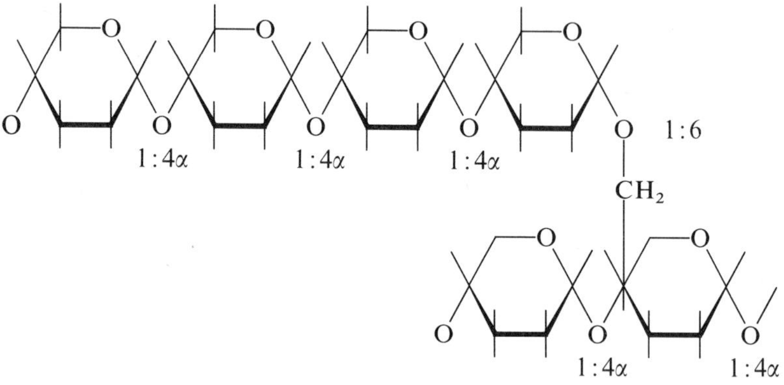

Part of an amylopectin molecule

Such a structure exists in another component of starch, amylopectin, which contains about 600 to 6000 glucose units (with molecular weights of 100000 to 1000000 or more). At least 50 points of branching are believed to be present in each molecule and we may represent part of the structure of amylopectin as shown in figure 1.3.

The main part of each chain is linked 1:4, whereas each point of branch is linked 1:6. Amylose and amylopectin together form the bulk of the material we describe as 'starch'; they are usually present in proportions of about 1:4.

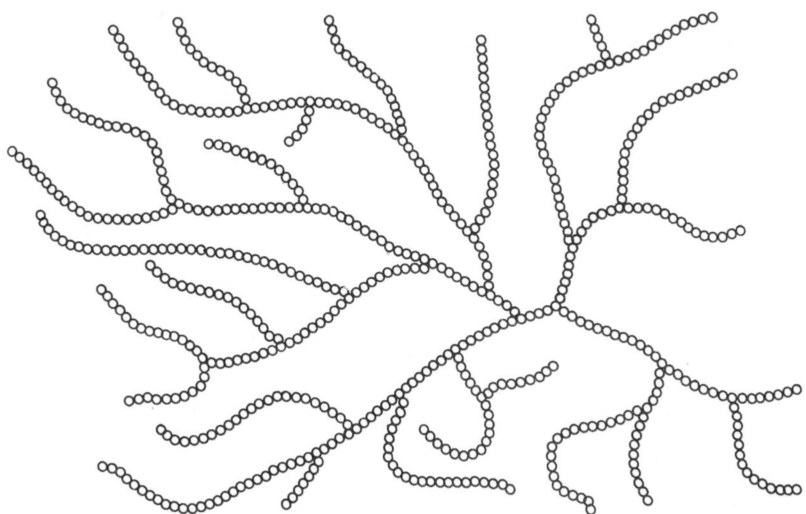

Figure 1.3
Diagram to illustrate the general structure of amylopectin. Each circle represents a glucose residue.
From Baldwin, E. (1967), The nature of biochemistry, 2nd edn, *Cambridge University Press.*

Glycogen, the storage polysaccharide of animals, is very similar in structure to amylopectin but the branching is thought to be more frequent, giving very short (12 glucose unit) chains, and its molecular weight is very large, probably several million.

Cellulose is a high molecular weight polymer formed from β-D-glucose. Cellulose molecules contain about 10000 β-D-glucose units (giving a molecular weight of about one million), and the packing together of large numbers of the threadlike cellulose molecules produces a material of great tensile strength. Cellulose is the chief structural carbohydrate of all plants. The diagram below shows simplified β-D-glucose rings linked together.

β-glucose β-glucose

CH_2OH CH_2OH

O O O O O O O O

CH_2OH CH_2OH

cellobiose unit

A simplified representation of β-D-glucose rings linked to form cellulose

The α-linkages can be broken by enzymes present in man, and the glucose units in amylose and amylopectin in starch can therefore be made available for use by the body. But the β-linkages are not broken by any enzymes present in man, and cellulose is of no nutritive value at all to humans. Horses and ruminants are able to digest cellulose in grass because enzymes produced by bacteria in their digestive tracts break down the β-links.

The body's main source of energy is the oxidation of glucose. Foodstuffs which are rich in polysaccharides are of value principally as sources of energy. After a foodstuff which provides glucose has been digested, the glucose passes in the blood to the liver and muscles where it is polymerized to glycogen, and where it remains until needed as a source of energy.

Vitamins

At the beginning of the twentieth century it was believed that a diet consisting of balanced quantities of fats, proteins, carbohydrates, mineral elements, and water would maintain health. However, experiments with animals revealed that they did not thrive on such a diet, and enquiries into the reasons led to the discovery of vitamins.

Most vitamins are fairly complex chemically, and they are quite different from each other in structure and function. Only small quantities are required by the body, but with few exceptions the body cannot synthesize them and if they are not provided in the diet deficiency diseases appear. Vitamin D, for example, controls the use and retention of calcium and phosphorus in the body, and is essential for bone and teeth formation. Thiamine, vitamin B_1, forms part of an enzyme system, and without it the enzyme cannot work.

Two vitamins are described here, the first because it is chemically simple, and the second because it is mentioned in several other parts of this book. One of the simplest known vitamins is nicotinic acid, which the body uses in the form of the amide, nicotinamide.

nicotinic acid

nicotinamide

Severe deficiency of nicotinic acid leads to the disease pellagra, which is characterized by dermatitis, diarrhoea, and symptoms of mental disorder. The vitamin is involved in a large number of oxidation processes in the body. Meat and fish contain ample quantities of the vitamin for human needs, and pellagra is usually found only in communities whose principal food is maize.

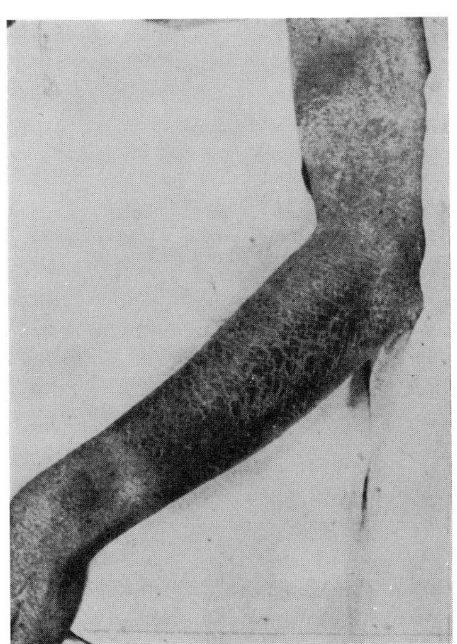
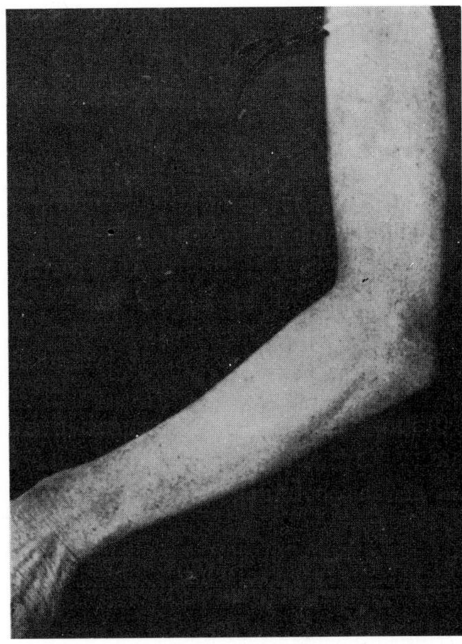

Figure 1.4
Vitamin deficiency – Pellagra (deficiency of nicotinic acid). The arm on the left shows the condition before treatment: brown pigmentation of skin, which is cracked and glossy. On the right is the same arm after a fortnight's treatment with a diet containing nicotinic acid.
From Simpson, S. L. (1935), *'Secondary Pellagra,'* Quart.J.Med. **28**, *191–201.*

Ascorbic acid, or vitamin C, was found to be the vitamin whose absence in the diet led to the disease scurvy. Scurvy used to be prevalent amongst sailors in the days of sailing ships, when crews were forced to spend long periods without fresh fruit or vegetables. The disease is characterized by haemorrhages under the skin, and swollen gums from which the teeth readily drop out. The prevention of the disease on long voyages by carrying citrus fruit or fruit juices was discovered long before the reason was fully understood. Citrus fruit juices are rich in vitamin C.

Ascorbic acid, in spite of its name, does not contain a free $—CO_2H$ group. The $—CO_2H$ group reacts with an —OH group in the molecule to eliminate a molecule of water and form a ring compound.

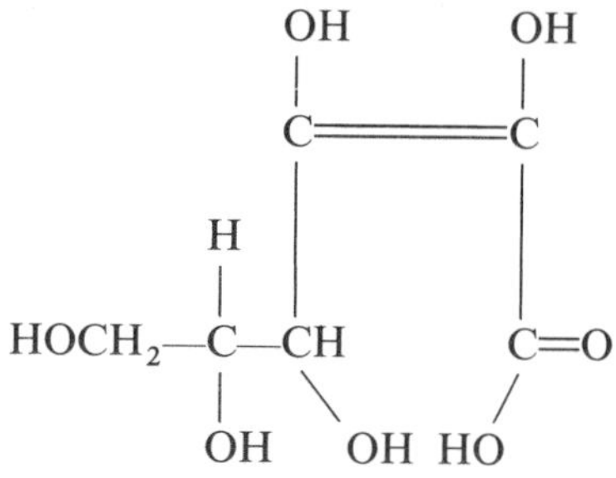

```
          OH        OH                       OH        OH
          |         |                        |         |
          C=========C                        C=========C
          |         |                        |         |
     H    |         |                   H    |         |
     |    |         |                   |    |         |
HOCH2—C—CH          C=O         HOCH2—C—CH            C=O
     |    \        /                    |     \       /
     OH    OH    HO                     OH      O
```

free acid corresponding to ascorbic acid

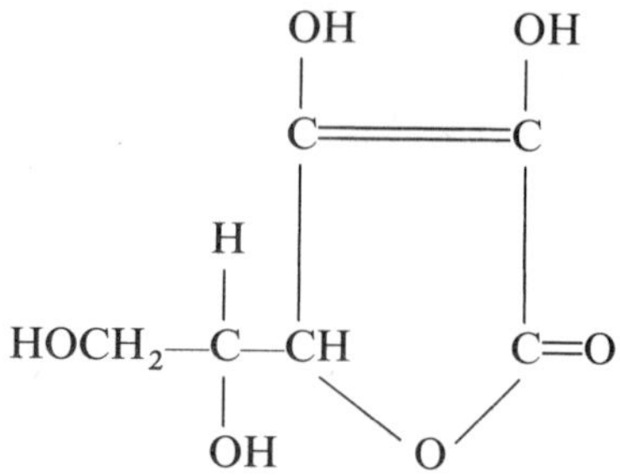

ascorbic acid

Ascorbic acid is a good reducing agent, and as such it is readily oxidized. Under certain conditions it may become oxidized even by exposure to air.

In the United Kingdom the principal sources of ascorbic acid are fresh vegetables, potatoes, and citrus fruits. Blackcurrants and rosehips are particularly rich sources.

Mineral elements

The term 'mineral elements' means elements other than carbon, hydrogen, oxygen, and nitrogen. The eight principal mineral elements in the body in terms of weight are calcium (about 1 kg in a man of average weight), phosphorus, potassium, sulphur, sodium, chlorine, magnesium, and iron (about 5 g in a man of average weight). In addition to these there are small quantities of other elements (trace elements), many of which are essential to life.

Calcium and phosphorus account for about three quarters of the weight of mineral elements. They are incorporated into the principal supporting material of the body – bones, and also into teeth, in the form of calcium phosphate. Calcium also occurs in the body fluids either as free ions or combined with proteins. The level of calcium ions in the blood must remain constant, and if not enough are present then calcium ions leave the bones and pass into solution in the blood. The calcium ions in bones and blood are in dynamic equilibrium. Phosphate groups form part of several classes of organic compounds, including the nucleic acids. The compound adenosine triphosphate (ATP) is essential for the functioning of muscles, providing a source of immediately available energy. Phosphoric acid and its salts are important buffer solutions in the blood.

Sodium ions and chloride ions are essential in the body fluids, and their concentration must remain constant.

Magnesium is involved in maintaining the balance of minerals in the skeleton, and it also activates several enzymes in muscle.

Most of the iron in the body occurs as haemoglobin, the compound by which oxygen is carried from the lungs to the tissues, and as myoglobin, which acts similarly in combining with oxygen in the muscles. A surprisingly small amount of the iron in food is actually absorbed by the body, 99 per cent of it being lost in the faeces. An adult male requires about 10 mg of iron a day, and a woman losing blood through menstruation will require about 12 mg a day. Iron is one of the elements which may be lacking in the diet of some people, and it is therefore added to all flour in the United Kingdom.

Haemoglobin can only be synthesized by the body, in the bone marrow, if small traces of copper are present in addition to the iron. Copper is also a component of several enzymes.

Other trace elements are zinc, manganese, and molybdenum, which are all concerned with enzyme action.

As many of these mineral elements occur only in some foods, it is essential that the diet should include a wide variety of foodstuffs.

Water

Analysis shows that all tissues of the body contain water, from 3 per cent in the enamel of teeth to 99 per cent in the cerebrospinal fluid. Water is the solvent in which the complex reactions of living processes occur. The movement of nutrients through the body is achieved by means of blood which is an aqueous solution and suspension. Water is also important in a wide range of other functions; for instance, it takes part in hydrolysis reactions, is an essential part of the structure of many plant and animal tissues, and maintains tissue rigidity especially in plants. Water is an exceptionally good solvent. Blood is about four-fifths water by weight and consists of dissolved ionic and molecular substances, such as simple sugars, together with substances in suspension, such as blood corpuscles. The hydrolysis of proteins to polypeptides and to amino acids involves the consumption of one water molecule at each of the split linkages, and the hydrolysis of polysaccharides likewise involves one water molecule at each split link.

In addition, the physical properties of water regulate the action of all tissues.

It can thus be seen that for the human body to build up its structure and for it to function properly it must obtain from food the ten essential amino acids, the necessary fatty acids, suitable carbohydrates, the vitamins, mineral elements, and, of course, water.

Human energy requirements

The body requires energy not only for heat and for movement but also to bring about synthetic chemical reactions, and to maintain the structure of its cells.

Needs for energy vary with such factors as age, sex, body weight, climate, physiological state, and the degree of physical activity. For an average man of 70 kg, of age 30, about 12 600 kJ per day are required for normal activity; the corresponding value for a woman is about 9200 kJ. A pregnant woman requires an additional 800 kJ per day, and one who is lactating an additional 2100 kJ per day. The number of kJ per hour expended in physical activity ranges from about 60 whilst sitting to 4000 in running.

Fats, proteins, and carbohydrates can all be oxidized by combination with oxygen, to provide energy. In complete combustion, fats produce approximately 36 kJ g^{-1}, proteins 16–20 kJ g^{-1}, and carbohydrates 16 kJ g^{-1}. Although it would seem that energy requirements could be most easily supplied by fat, there are limits to the amount of fat which the body can cope with, and since fat is not soluble in water it cannot be readily transported by the blood to where the energy is required. Sugars are soluble in water and may be rapidly transported by the blood.

The assimilation of nutrients

It is not enough to be given foods containing the nutrients which the body needs in the required quantities. They must be taken into the body cells by absorption from the central digestive tract. The walls of the stomach and intestines are equipped with specialized cells which discharge enzymes into the food eaten and which can digest it into a suitable form for entry into the cells of the body. In the stomach, for instance, under the action of hydrochloric acid and the enzymes in gastric juice, proteins are broken down to smaller relatively soluble molecules, and some breakdown of fats to their component fatty acids and glycerol occurs.

Further digestion and absorption take place in the small intestine. In the large intestine a further absorption of nutrients from the partly digested food mass also occurs but the main importance of the large intestine is reabsorption of water and minerals which have been secreted into the food as digestive juices.

Biochemical individuality

Although the types and quantities of food requirements for human beings have been scientifically investigated, the data represent average values. Between normal, healthy individuals there are very wide ranges in body composition and structure, and in enzymic constitut on. The variation in stomach volume is eight-fold, in susceptibility to poison a hundred-fold, and in peptic activity four thousand-fold. Again there are very great differences in the size of individuals' livers, and in the shape and position of the major blood vessels which distribute blood (and nutrients) to the different parts of the body. Even in the range of responses to a taste sensation as common as bitterness there are differences in the reactions of perfectly normal, healthy people. For example, one person in four cannot detect phenylthiocarbamide, a simple chemical which is intensely bitter to three out of four people.

There are eight-fold differences between normal individuals in the requirements for amino acids. A four-fold difference in vitamin C requirements has also been shown. Many other examples could be given.

It seems likely that many of us are suffering from slight individual nutritional deficiencies of which we are quite unaware. The wide variation of the age of onset of senility and of organic disease suggests that this is not idle speculation; for instance, substantial groups of people live active lives until they are about 100 years old in Georgia in the USSR. It is clear that the average person can hope for a longer healthy life when the causes of such biochemical individuality have been discovered.

Chapter 2

The texture of food and its quality

If a man were provided with bottles containing a range of isolated nutrients such as fats, the essential amino acids, glucose, vitamins, and soluble salts of mineral elements, he could make up a mixture which when taken with water would enable him to live. But such a mixture would be most unpalatable.

We enjoy food not only because of its nutritional value but also because of the quality of the food: the crispness of fresh fruit and well cooked vegetables, the tenderness and juiciness of meat, and the taste and odour. Our food is normally not just a mixture of pure chemicals; it is part of the structural and fleshy materials of plants and animals, and the chemicals which constitute these are not a random mixture, but are arranged in a very precisely ordered fashion for performing essential functions in the living organisms. It is these ordered arrangements which give quality to natural foods.

It is convenient to distinguish two principal categories of eating quality. One is the structural character: texture, water holding capacity, tenderness, and juiciness. The other is chemical character: taste, odour, and colour. The distinction is not wholly logical since the structural attributes are due to ordered arrangements of chemicals, but the classification is a useful one.

Plant and animal tissues are composed of discrete units, or cells, and each cell is bounded by a cell membrane. The membrane is a layered structure made up of lipids (the class of compounds of which fats are one group) and proteins; the material is therefore known as lipoprotein. Figure 2.1 shows the arrangement.

The polar groups of the lipids are attached by electrostatic attraction to the polar groups in the protein, forming a layer. The long hydrocarbon tails of the lipids in one layer intermingle in an orderly way with the hydrocarbon tails of another layer, forming a double layer with the hydrocarbon portions of the lipids in the centre.

This lipoprotein arrangement proves to be selectively permeable to ions and molecules, and it forms the semipermeable membrane by which osmotic phenomena can operate in the cell.

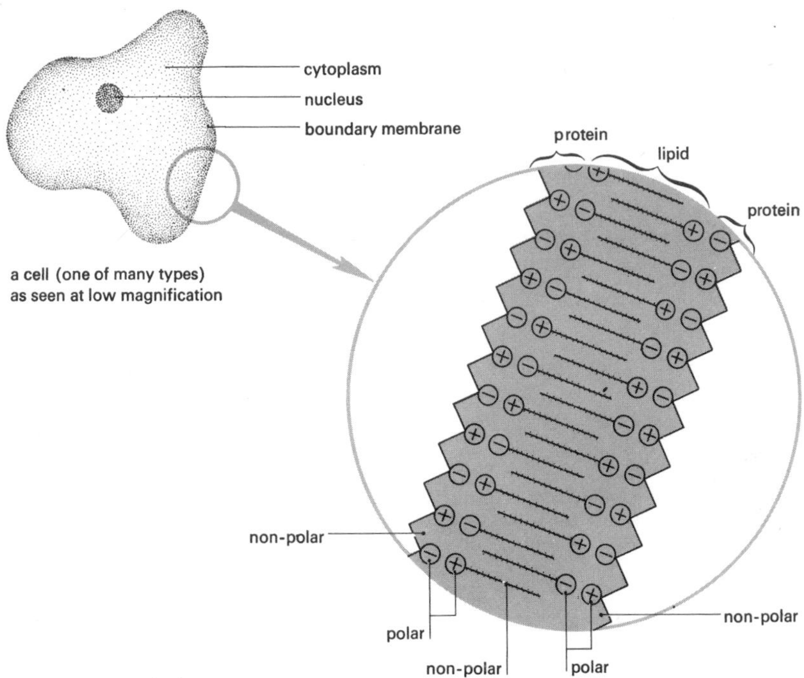

Figure 2.1
A diagrammatic representation of the chemical structure of a cell membrane. At the very high magnification achieved by electron microscopes the membrane is seen to be a double layer; each part of the layer is composed of lipid and electrostatically adsorbed protein.

Water holding capacity

Since water is the major component of living cells, it is also the principal constituent of foods, and a substantial part of the texture, tenderness, and juiciness of foods is determined by the various ways in which water is arranged in relation to insoluble materials. A small proportion of the total water is chemically bound to the polar groups of proteins and polysaccharides. This is achieved by hydrogen bonding between water molecules and the N—H and C═O groups in proteins, and the —OH groups in the polysaccharides. In addition, some molecules of water are held by the hydration of free ions or of ions such as Mg^{2+} and Ca^{2+} which occur in some protein structures. But a much larger proportion of water is non-combined or 'free', with properties the same as those of liquid water outside the living organism. Since water does not pour out of plant or meat commodities it is clear that this non-combined water must be held immobile in some way. It is believed that the long chains of the proteins link with one another by means of hydrogen bonds and ionic bonds to form a meshwork, and water is trapped in the interstices. Long chain polysaccharides also link up by means of hydrogen bonds to form an entrapping meshwork. Meshwork structures such as these which have trapped large quantities of water are known as *gels*.

Any factors which affect the bond forces holding the protein or polysaccharide chains together will affect the extent to which water can be held by the system. One factor is pH, which affects the electrostatic charges on the protein molecules. Another is heating, which ruptures hydrogen bonds. In these ways the meshwork may break down.

Acids, alkalis, and high temperatures, may affect not only the meshwork structure but also the tertiary and secondary (helical) structures of the individual protein molecules. Such denaturation will usually lead to coagulation, solidification, and a loss of water holding capacity.

A fresh strawberry, although it has a high total water content, is firm enough to offer immediate resistance to the cutting edges of the teeth. This firm 'feel' and the sensation of juice being released from the cells as the berry is bitten adds greatly to the enjoyment of eating fresh strawberries. A frozen strawberry is a quite different article. The cell structure has been disrupted by freezing and thawing, and much more water is in the intercellular regions. This drips out of the thawed berry and the berry has a soggy texture. It offers no resistance to the teeth until it has completely collapsed; the sensation in the mouth is quite different from a fresh strawberry, and some people dislike it.

Texture in foods of plant origin

In plants, carbohydrates perform a role of major importance both as structural materials and by reason of their water holding capacity. The cell membrane is surrounded by a cell wall which imparts protection and firmness. The cell membrane is composed of lipoprotein (figure 2.1), and is semi-permeable. The cell wall is largely composed of carbohydrates, and it is usually freely permeable to solutes. It is the cell wall and its constituent polysaccharides which are primarily responsible for the texture of plant foods. Figure 2.2 shows diagrammatically and in simplified form the structure of two adjacent cell walls of a plant.

The *hemicelluloses* form a group which is difficult to define and to isolate. On hydrolysis they give a variety of different hexose sugars (for instance glucose and mannose), pentose sugars, and uronic acids. These components appear in widely differing relative amounts and in numerous combinations, reflecting a bewildering variety of mixed polymers.

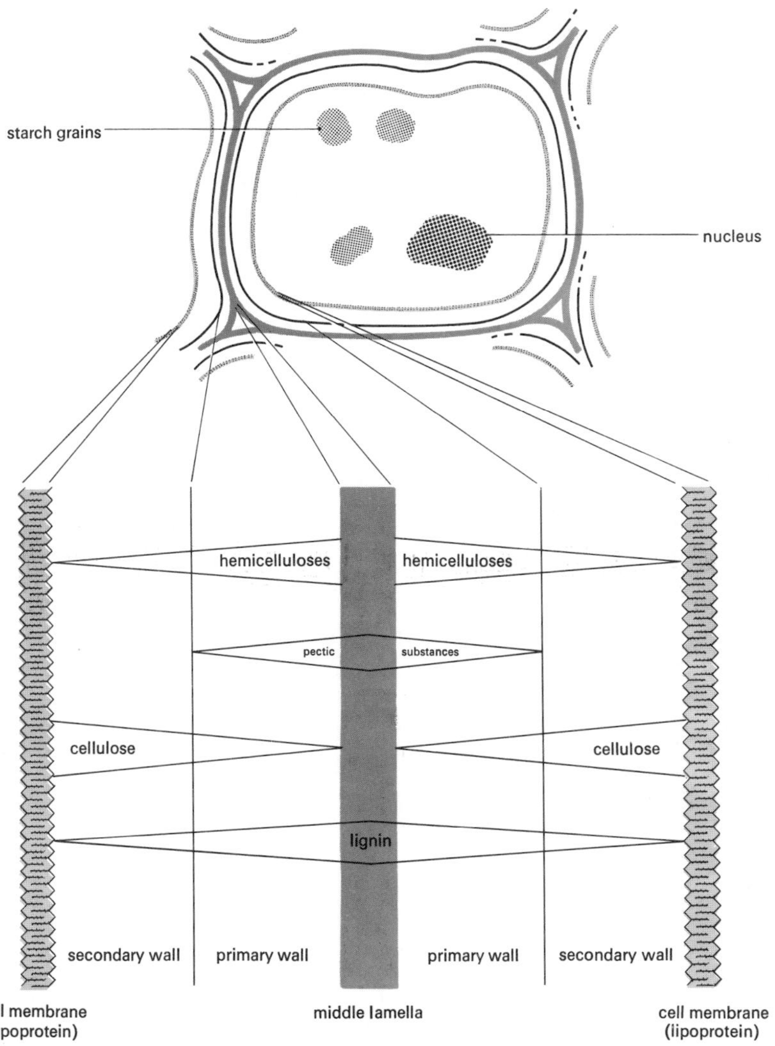

Figure 2.2
A much simplified diagram of two adjacent plant cell walls and their principal chemical constituents. (Not to scale.) The wedges indicate the directions in which the substances become more, or less, prominent.

Pectic substances are also difficult to isolate. An aqueous extract yields pectin which is a mixed polymer whose monomers are α-D-galacturonic acid and the methyl ester of this acid.

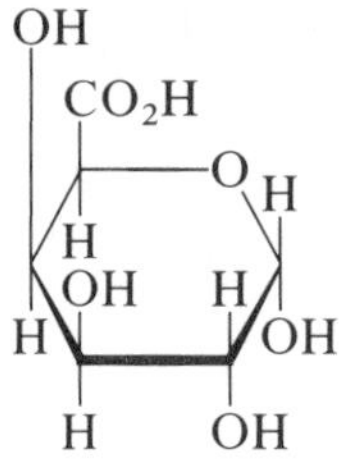

α-D-galacturonic acid

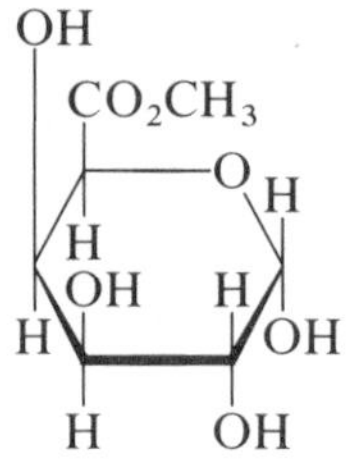

methyl α-D-galacturonate

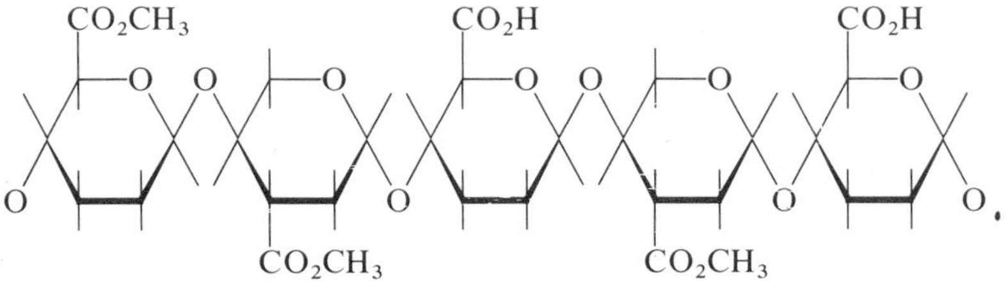

Part of a pectin molecule (methyl pectate)

The number of units in the chains varies with the type of plant material, with its age, and also with the method by which the pectin has been extracted from the plant. It ranges between 50 and 2000 (giving molecular weights of between 10000 and 400000). The middle lamella is largely composed of pectic material, some of it probably in the form of its calcium salts. The calcium content of cooking water can, by cross-linking pectin molecules together, affect the texture of cooked vegetables.

Pectins are important in the setting of jams. Firm jams and 'runny' jams may have little or no difference in water content. The important difference is in the amount and state of the pectin in the fruit after boiling the jam. This determines the firmness of the gel structure in the jam after cooling.

Attempts have been made to improve frozen strawberries by adding pectin preparations so that the juice sets to a jelly instead of running freely. However, although the juice does not run out of the berry, the sensation is no different from the untreated frozen berries when they are bitten into.

Lignin is not a polysaccharide. Pure lignin is not easy to isolate and for this reason its structure is not yet known for certain. It appears, however, that it is a polymer of a derivative of phenylpropane.

$$\left[HO\text{—}\bigcirc\text{—}CH_2CH_2CH_2\text{—} \atop CH_3O \right]_n$$

The hardness of all wood, nut shells, and the toughness of celery fibres arise from the presence of lignin.

Cellulose is the principal substance giving strength to plant tissues. Its chemical structure, a polymer of β-D-glucose, was described in chapter 1. X-ray diffraction studies of cellulose fibres show that they contain crystalline areas. In these the cellulose molecules are arranged parallel to each other, and are held in this ordered way by hydrogen bonding. In the disordered, amorphous areas there is a greater number of free —OH groups than in the crystalline areas, and it is mainly in the amorphous areas that absorption of water takes place, by hydrogen bonding to these free —OH groups. Also, the amorphous part of the fibre is believed to be flexible. If this is so, the dual structure of cellulose is of especial value to the plant, the crystalline regions conferring strength, and the amorphous regions conferring flexibility.

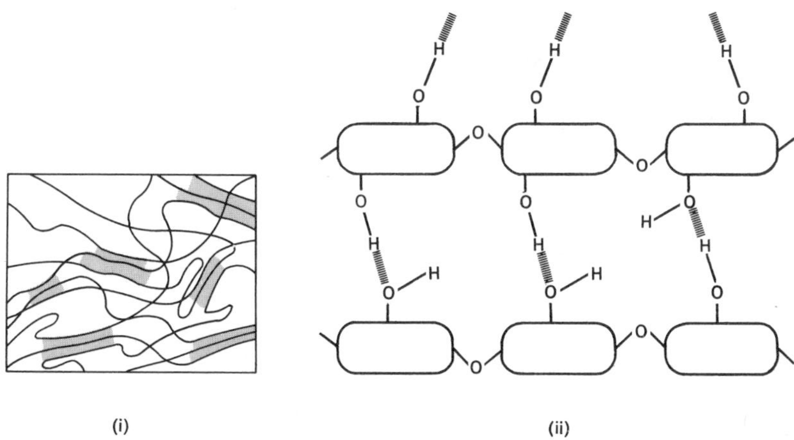

Figure 2.3
Crystallinity in cellulose. (*i*) Ordered crystalline regions in cellulose fibres. (*ii*) Hydrogen bonding between cellulose molecules to produce ordered regions.

The water holding capacity of plant cells is governed principally by three features: the lipoprotein cell membrane, the carbohydrates in the cell wall, and the proteins and carbohydrates within the cell. The lipoprotein double layer is, as has been described earlier, a semipermeable membrane, and controls the movements of solvent and solutes into and out of the cell by osmosis. The carbohydrates bind water by hydrogen bonding, and proteins do so by a combination of hydrogen bonding and the hydration of ions in the protein chains.

The water holding capacity of pectins varies with the degree of esterification of the $—CO_2H$ groups, and that of cellulose with the extent of the crystallinity. Since heating leads to a rupture of hydrogen bonds, boiling cellulose in water will lead to some breakdown of the crystalline regions and an increase in water holding capacity. This is one of many reasons for textural changes in cooking.

This phenomenon is particularly marked in the swelling of starch grains when these are heated in water. The increased absorption of water causes the starch cell to burst, the viscosity of the starch–water mixture increases owing to the establishment of a loose three-dimensional structure involving amylose chains and water molecules hydrogen bonded together, and a gel may form.

Texture in foods of animal origin

Unlike plant cells, those of animal tissue have no walls of cellulose or hemicellulose outside the bounding double membrane of lipoprotein. The main structural elements of animal cells consist of protein polymers.

The edible portions of animals include connective tissue, which consists substantially of fibres of collagen with a much smaller number of fibres of elastin. The fibres are cemented together by an amorphous matrix. The properties of connective tissue derive, of course, from the chemistry of its constituent proteins. *Collagen* is found in varying forms representing degrees of aggregation of the fundamental threadlike molecules. The latter consist of three non-identical polypeptide chains. Approximately a third of the amino acid residues is glycine. The individual chains each form a lefthand helix, and they are intertwined to form a righthand super helix. This structure is stabilized by hydrogen bonds. The molecules aggregate to form fibrils and these, in turn, aggregate, eventually producing fibres visible under the light microscope.

There is an increased formation of cross bonds between the individual polypeptide chains in each molecule, and especially between neighbouring molecules in the fibrils, with increasing animal age. In addition to these 'qualitative' changes, the quantity of collagen in connective tissue increases with animal age. These changes help to explain why the tissues of older animals are more difficult to chew than those from younger animals.

When the temperature of collagenous connective tissue is raised in cooking, there are changes in molecular structure which make chewing easier. There is some shrinkage between about 60 to 64 °C, and above 100 °C collagen is converted into a soluble degradation product, gelatin.

Elastin, the other principal protein found in connective tissue, consists fundamentally of polypeptide chains randomly linked, and cross-linked at intervals by strong covalent bonds. The tendency to form an ordered α-helix between the cross-links is severely restricted, because of the rather large numbers of proline and hydroxyproline amino acid units in the chains.

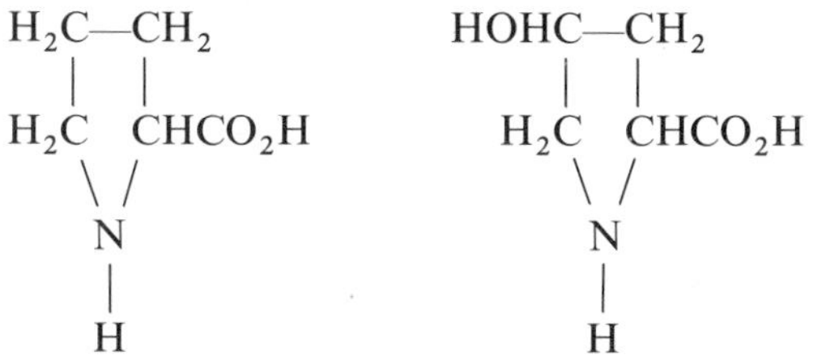

$$\begin{array}{ccc} H_2C\text{—}CH_2 & \quad & HOHC\text{—}CH_2 \\ |\quad\;\; | & & |\quad\;\; | \\ H_2C\quad CHCO_2H & & H_2C\quad CHCO_2H \\ \backslash\;\; / & & \backslash\;\; / \\ N & & N \\ | & & | \\ H & & H \end{array}$$

proline hydroxyproline

You may like to consider why the presence of these two amino acid units in a protein would hinder the formation of an α-helix. When the molecules link up with other amino acids to form a chain, how many hydrogen atoms remain attached to the nitrogen atoms in the proline and hydroxyproline molecules? What result would this have?

Up to 70 per cent of the total edible tissue of animals is muscle, so its characteristic structural elements, *myofibrils*, must clearly be considered in relation to eating quality. The structure of muscle at different magnifications is represented in figure 2.4.

Myofibrils are composed of longitudinally-repeating units, sarcomeres, which embody the basic mechanism of contraction. The sarcomere is bounded at each end by transverse divisions, the Z-lines. Its central region contains a parallel array of thick filaments of the protein, *myosin* (which constitutes about 75 per cent of the myofibril). Extending towards the centre of the sarcomere from the Z-line on each side are thin, parallel filaments of the protein, *actin*. These penetrate between the myosin filaments.

According to the generally accepted 'sliding filament' scheme, chemical interaction between the myosin and actin filaments during contraction causes the latter to pull the Z-line from each side of the sarcomere. Hence the degree of

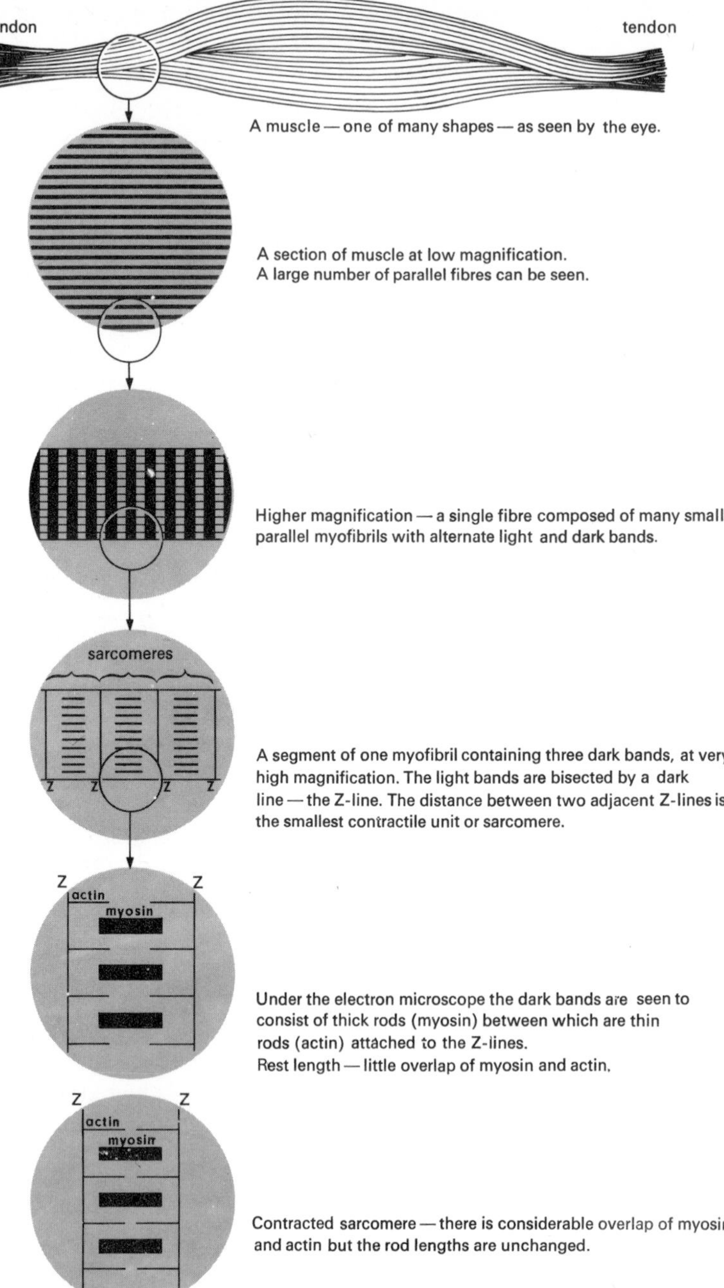

Figure 2.4
The structure of meat (muscle) at different magnifications.

contraction observed in a muscle directly reflects the degree of overlap of myosin and actin filaments.

From the point of view of eating quality, the importance of contraction is that it may occur after slaughter, to a greater or lesser extent according to circumstances, and that the degree of contraction contributes directly to the overall degree of toughness to the palate. It has been suggested that this is because myosin filaments from adjoining sarcomeres will tend to form bonds with one another through the Z-lines.

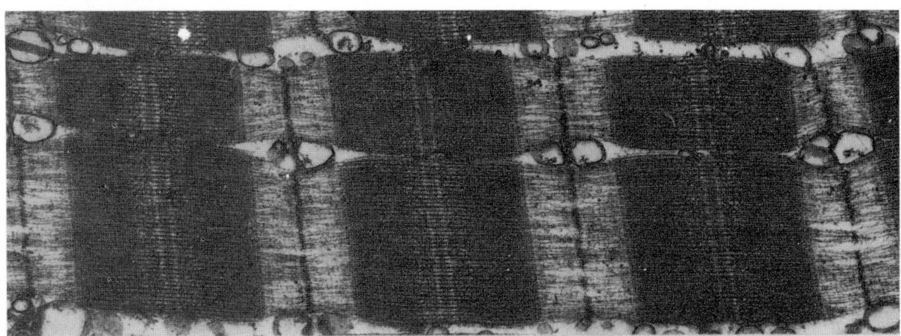

Figure 2.5
An electron micrograph showing a section of two myofibrils, each containing three sarcomeres (×20000). The sarcomeres are bounded on each side by the dark Z-lines which bisect the light I-bands. The magnification corresponds approximately to the fourth sketch in figure 2.4.
Photo, Unilever Research Laboratory.

In cooking meat, considerable contraction can be expected; nevertheless the degree of shortening before cooking significantly affects the tenderness.

Apart from the possibility of contraction after slaughter, another factor peculiar to muscular tissue which affects texture and tenderness, is the onset of rigor mortis. This shows as a stiffening of the tissue because in the dying tissue the actin and myosin filaments cross-link irreversibly. If meat is cooked before the onset of rigor mortis it is more tender.

Flavour and colour

Of all the attributes of eating quality, flavour, a term which embraces both taste and odour, is by far the most complex. The related sensations vary greatly in quality and quantity, and even when a flavour has been precisely defined in chemical components, the reaction of the individual consumer is most variable for both physiological and psychological reasons (sex, age, body-type, temperament, association).

The quality of taste is exclusively detected by receptors in the mouth, and the substance to be tasted must be in solution. Taste involves four basic sensations: salt, acid (or sour), sweet, and bitter. The first two are most keenly appreciated at the sides of the tongue, whereas sweetness and bitterness are detected particularly at the tip and back respectively.

Acidity and salinity can be related fairly closely to chemical structure, as they depend on the presence of ions – the hydrogen ion in the case of acidity. The sensations of sweetness and bitterness cannot be related readily to chemical constitution. Such apparently unrelated structures as lead nitrate, glucose, and saccharin are sweet, and such diverse compounds as picric acid, quinine, and magnesium sulphate are bitter.

The quality of odour is detected by receptors in areas at the top of the two nasal cavities. Volatile substances soluble in water or fat, or gaseous suspensions of insoluble particles, can be responsible. Although attempts have been made to classify odours as fragrant, spicy, ethereal, resinous, acid, burnt, caprylic, and foul, the sensations are as varied as the substances which arouse them.

The nose is an extraordinarily sensitive detector of odour; it can beat the most modern analytical instrument for detecting trace amounts of odorous material, and can sense substances such as vanillin at concentrations as low as one part in 50000 million parts of air.

The reason why different compounds have different odours is still unknown, but it appears to be related to the size, shape, and charge distribution of the molecule.

The senses of taste and smell, in addition to the pleasure they can give, are useful as a defence, in that they allow people to detect and avoid food spoiled by micro-organisms.

The flavours by which consumers characterize foods are due to the interaction of taste and odour. In many cases the substances responsible are not originally present in the food. They are deliberately fostered by holding the commodity for some time in conditions which induce beneficial alterations (for instance, the holding of fruit to ripen, and of meat to induce tenderness).

The methods of processing used to preserve foods may alter flavours in both desirable and undesirable directions. Thus the flavour of canned salmon is preferred by many people to that of fresh; canned strawberries are justifiably regarded as different from fresh strawberries in flavour, texture, and colour;

and the odour of meat irradiated by γ-rays to the point of sterility is much less attractive than that of fresh meat.

Quite apart from variation in ability to taste, the presence of the component responsible for a given flavour sensation can be masked deliberately or unintentionally by chemical, physical, and psychological means. Thus the 'off' odours in meat can be disguised by the use of sufficient seasonings during cooking. Also, if the structure of the commodity is 'close' rather than 'open', components may not be released sufficiently quickly during mastication for detection by the palate (for instance, bacon of high pH does not taste so salt as that of low pH, for a given salt content). If a flavour is presented out of its familiar context, the consumer's response can be affected.

However important the colours of foods may be to the eater, the compounds which give rise to them are important to plants and animals for physical or chemical reasons and usually not for the colours themselves. The effect of their pigmentation on the human eye is usually quite irrelevant.

In food of plant origin, three groups of compounds are principally responsible for colour: the porphyrins, the carotenoids, and the flavonoids. The porphyrins are large ring-shaped compounds with a metal atom at the centre of each molecular ring. Chlorophyll is a porphyrin, the metal atom being magnesium. The ability of plants to absorb light energy for the synthesis of carbohydrate from carbon dioxide and water is due to chlorophyll. The carotenoids are usually yellow in colour; their function is not known but they may be important in photosynthesis. The flavonoids are polyphenols and they are responsible for the red and blue colours in fruit and vegetables, for instance the red of beetroot and of oranges.

In foods of animal origin, the porphyrins are mainly responsible for colour. The metal atom at the centre of the porphyrin ring is iron. The red colours of blood and meat, and the brown colour of hens' eggs are due to iron porphyrins.

Measuring the attributes of food quality

In order to find the most efficient means of producing good quality food methods of measuring food quality quantitatively are necessary.

Subjective assessment of eating quality by taste panels has long been employed. Even with the development of sophisticated, accurate objective measurements, the importance of the taste panel remains since it is needed to calibrate instrumental findings against human reactions. Taste panels of both trained and untrained people are used. The former are able to detect specific changes in texture, juiciness, colour, and flavour in controlled experiments; the latter give the spontaneous reaction of the consumer to the products. Some commodities, especially those which yield alkaloidal and alcoholic beverages, are assessed by experts.

In taste panel operations, several methods are used to express the intensity of the parameter being assessed. In some, numerical values are given to the attribute; in others, less precise evaluations are made. Samples may be assessed on pleasure-giving or numerical scales of preference; or a triangle test may be used in which panel members attempt to pick out that sample, in a group of three, which alone possesses or does not possess a particular attribute.

In assessing flavour, attempts have been made to match the substance against a limited number of descriptive terms the interpretation of which is strictly defined and understood. This is the 'flavour profile' method.

Machines used for the objective measurement of texture are designed to measure a variety of mechanical properties of the food. Some machines measure the force needed for a blade to shear through food (for example the 'tenderometer' used to measure the tenderness of peas, see figure 5.1), or the force needed to thrust a thin cylinder through food (the basis of various 'penetrometers'), or response to static or dynamic deforming forces. None of these gives completely satisfactory correlation with the subjective human assessment of texture, because the chewing action, which gives the sensation of texture, is a complicated mixture of cutting, shearing, crushing, and grinding movements, and cannot be represented by just one of these movements. Attempts have been made to develop more realistic machines, even machinery incorporating false teeth or an 'artificial mouth', but the final assessment must still be left to the human taster.

Since the intractable materials characteristic of plant and animal foods are cellulose and collagen respectively, chemical determination of the breakdown product is sometimes used as a measure of textured properties in foods. This can be unreliable because the most important feature in determining meat texture is the length, thickness, and orientation of collagen fibres rather than the total amount. Thus a steak can be made more tender by beating it before cooking, but analysis would show that it contained the same amount of collagen after the beating as before.

No suitable method is yet available for measuring the quality of flavour, although very sophisticated equipment has been used for separating and identifying the individual aroma components from foods. Concentration has been effected by freezing, zone-refining, and short-path vacuum distillation.

Separation and detection have been especially elegantly achieved by gas chromatography. Using a flame-ionization detector, 10^{-12} g of a compound separated by the gas–liquid column can be registered.

Flavour-giving components can now be separated and identified by many different methods of chromatography (on paper, on thin layers of cellulose supported on glass plates, and on solid–liquid, liquid–liquid, and liquid–gas columns), and by electrophoresis. Components may also be identified by infra-red and mass spectrometry, and by nuclear magnetic resonance. However, none of these yet match the human nose for sensitivity and discrimination.

Figure 2.6
Apparatus under development for the infra-red analysis of milk.
Photo, National Institute for Research in Dairying.

Objective means of assessing colour depend either upon extraction of pigment and its estimation by spectrophotometric devices, or upon determination *in situ* by comparison with standard international colours, or by spectrophotometric measurement.

Other attributes of food

Certain other factors, even less tangible than taste, texture, and colour, govern the acceptability of food. Religious beliefs play an important role; for example, pork is not eaten by orthodox Jews, or beef by Hindus.

There is very often a great resistance to accepting unfamiliar food materials. Extreme examples have occurred during famines in areas where rice is the staple diet. When the authorities have sent supplies of wheat to these regions the local inhabitants have been reluctant to believe that it is edible and, even when persuaded to try it, they have used the traditional methods for cooking rice, by boiling the wheat grains in water, and the result has been practically inedible. Such people have starved to death surrounded by sacks of wheat grains.

If you think that your own choice of food is free from social and psychological prejudice, you are deluding yourself. Would you eat your dinner with pleasure if you knew that the meat was rat or cat, or dog or horse? Many people would find it quite impossible to eat these knowingly. Similarly you may refuse food for no better reason than because the smell or appearance reminds you of something unpleasant.

Chapter 3

Changes in food

1 Introduction

In the previous chapters we have discussed food in terms of its chemical composition and physical structure. Most emphasis has been on 'fresh' or 'raw' food, that is food immediately available after harvest or slaughter. At this stage food is the end product of the agricultural, horticultural, and fishing industries. Raw food has very serious limitations and shortcomings. Food has to be transported from the locality where it is grown to the place where it is to be eaten. Demand for food is constant throughout the year, but most agricultural production is seasonal and food must be in a form that can be stored and transported without deterioration in quality.

What problems arise in the storage of food? We have referred to fresh raw food as the end product of the agricultural industry, but it is very different from the end product of most other industries. If you are in the steel industry producing nuts and bolts, or in the plastics industry producing expanded polystyrene, or in the mining industry producing coal, the end product, unless grossly mishandled, will keep its original condition and quality for several years. In contrast the products of agricultural industry deteriorate on storage, often quite rapidly. The table below gives the length of time certain foods will keep before showing obvious signs of deterioration, assuming storage in dry conditions at 10 to 15 °C.

wheat grains	several years
potatoes	6 to 9 months
raw meat	2 days
strawberries	1 to 2 days
green peas (mechanically podded)	4 to 6 hours

Much deterioration in quality is due to microbiological spoilage caused by bacteria, yeast, or fungi. Food which is suitable for us is unfortunately suitable also for micro-organisms, and mankind is continually competing with micro-organisms for available nutrients. Many micro-organisms spoil the palatability and attractiveness of food, and some are also a danger to health. However, we must not forget that man uses some micro-organisms for his own ends, for example in cheesemaking, wine manufacture, and brewing.

Even changes which, in time, make the food unpalatable and harmful may be considered desirable in their early stages. It is a matter of opinion. Liking for flavours, for example, varies considerably between countries and between individuals within a country; some people prefer their butter rancid rather than fresh; other people like strong tasting meat. The desirability or otherwise of a reaction in food also depends on the required end products. If you want to make wine, breakdown of grape sugars into alcohol by micro-organisms is essential, but if you want to store grapes as fresh fruit this microbial action is highly undesirable.

Micro-organisms and food spoilage

Micro-organisms are everywhere: they are found in air, in water, and in soil, in hot springs at a temperature of 80 °C, and in the sub-zero temperatures of Antarctic ice. It is certain that any surface, unless deliberately sterilized and protected, will harbour some micro-organisms. The animals and plants which become our food all carry surface micro-organisms. Human beings are no exception. There are normally between 1000 and 10000 bacteria per square centimetre on the skin of the hands. Even higher populations per unit area can be found in the mouth, throat, lungs, and intestine. Most of these bacteria are quite harmless to their host. Man can co-exist with the great majority of his resident bacteria, but it is another matter if he has to share his food with them. All food, when it is in a live condition prior to slaughter or harvest, carries bacteria, but only on the surfaces such as the skin and intestine. Any bacteria which invade the tissues are dealt with by the normal defence mechanism of the plant or animal, such as the leucocytes in mammals. After death these defences cease to function and in time bacteria invade the interior of the food.

We have to accept the fact that all raw food will be contaminated by micro-organisms, the majority on the surface of the food. The main types of micro-organisms found on foods are moulds, yeasts, and bacteria.

Moulds are the largest species of food spoilage organism. They are a microscopic form of fungi, and consist of filaments of cells which join up to form a network visible as mould on foods, particularly on foods stored in slightly damp conditions. Moulds are almost always undesirable, although they contribute to the flavour of blue and green cheeses.

Yeasts are another group of fungi, and exist as oval shaped cells about seven microns long. They cause spoilage in some food but are also of considerable importance in the food industry. Many yeasts can convert sugars into alcohol and gaseous carbon dioxide, which is the basis of the production of alcoholic beverages. The carbon dioxide production is utilized for leavening dough in bread making.

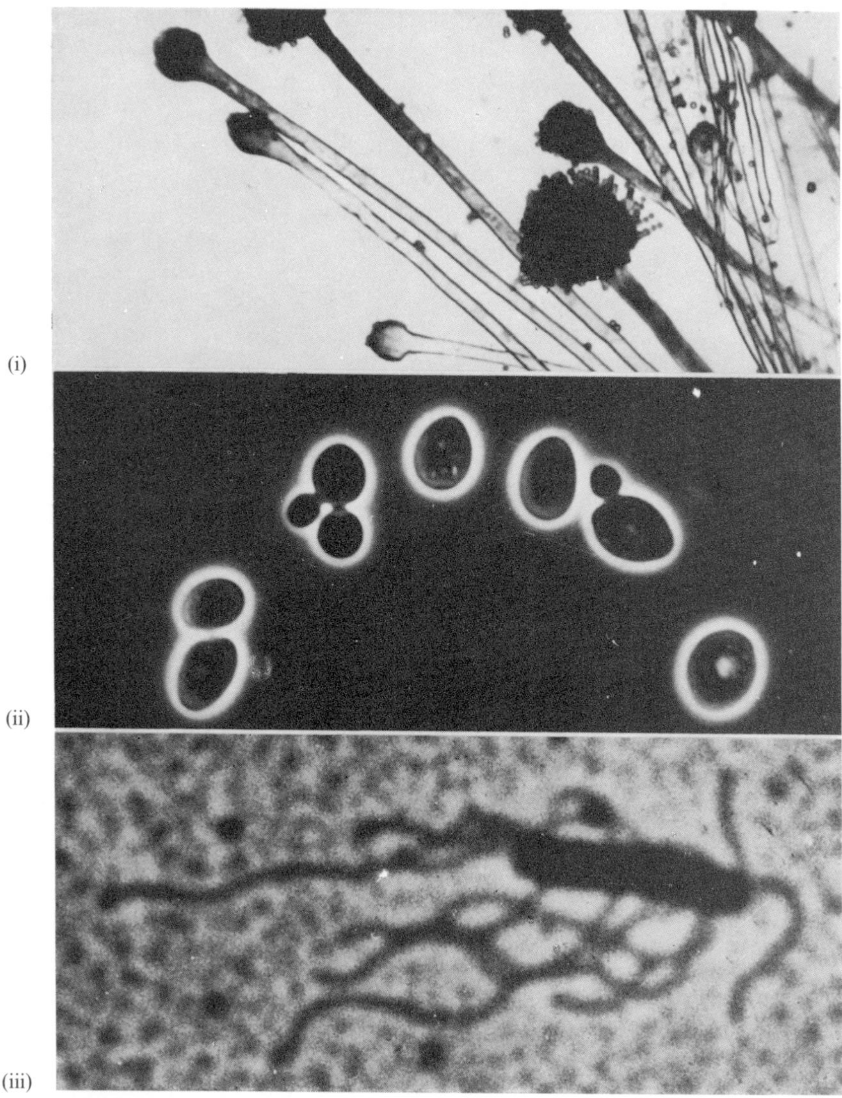

Figure 3.1
Types of food spoilage micro-organisms. (*i*) Fungus: A photomicrograph of mycelium of *Aspergillus flavus* (×50). (*ii*) Yeast: A phase contrast photomicrograph of cells of *Saccharomyces cerevisiae*, showing some cells in process of division (×2000). (*iii*) Bacterium: A photomicrograph of *Salmonella typhimurium* showing cell and flagella (×10000).
Photos, Unilever Research Laboratory.

Bacterial cells are smaller than yeasts, about one to three microns in size. Two particularly common types are the cocci which are spherical and the bacilli which are rod-shaped. Undesirable bacteria can be classified into two classes, spoilage organisms and pathogens. Spoilage organisms make food unpalatable and reduce the nutritive value but do not in themselves present any danger to health. Typical examples are *Lactobacilli* which produce lactic acid and cause souring in foods, and some *Clostridia* which break down proteins to give sulphur- or amine-containing compounds with putrid smells.

Pathogens cause illness or death in man. Typical food pathogens are:

Salmonellae. These organisms are carried on animals or immune humans. They are infectious in man and cause enteritis. About 5000 cases of illness due to *Salmonellae* are notified in Great Britain each year, but very few cases are fatal. One variety of *Salmonellae* causes typhoid. It not only causes enteritis, but also invades other organs and tissues and, until recently, this disease was often fatal.

Staphylococcus aureus. This organism is found on animals and is carried by most humans, particularly on the skin and in the nose and throat. The organism itself does not cause illness directly, but it produces a poisonous substance (enterotoxin) which irritates the stomach and intestine. The bacteria are killed relatively easily by heat but the toxin is heat stable, and consequently, if infected food is re-heated, the bacteria are killed but the toxins remain. It often occurs in cold food which is handled and then left to stand at room temperature before it is eaten, and sometimes affects a spectacular number of victims who have eaten a cold buffet at a civic banquet or wedding breakfast. About 500 cases occur each year in Great Britain.

Clostridium botulinum. This organism is normally soil-borne but some strains are water-borne, and vegetables, meat, or fish may be contaminated. *Clostridia* are spore-forming organisms which, when conditions are adverse, form spores with a very tough coat, and remain in a more or less dormant condition until conditions become more favourable for growth. Spores are much more resistant to heat than most micro-organisms. In its vegetative (growing) condition *Clostridium botulinum* makes a toxin which is probably the most poisonous substance known. It is a neurotoxin which affects the central nervous system and is usually fatal. Luckily the toxin is destroyed by heating even though the bacterial spores may survive. Cases of *botulinum* poisoning are usually caused by canned or bottled vegetables which have been insufficiently heated. No case has been reported in Great Britain for several years.

Clostridium welchii. This is generally similar to *Clostridium botulinum* but the effects are very much milder. It requires a long time for growth and usually occurs in stews which are heated and then left overnight. About 4000 cases occur each year in Great Britain.

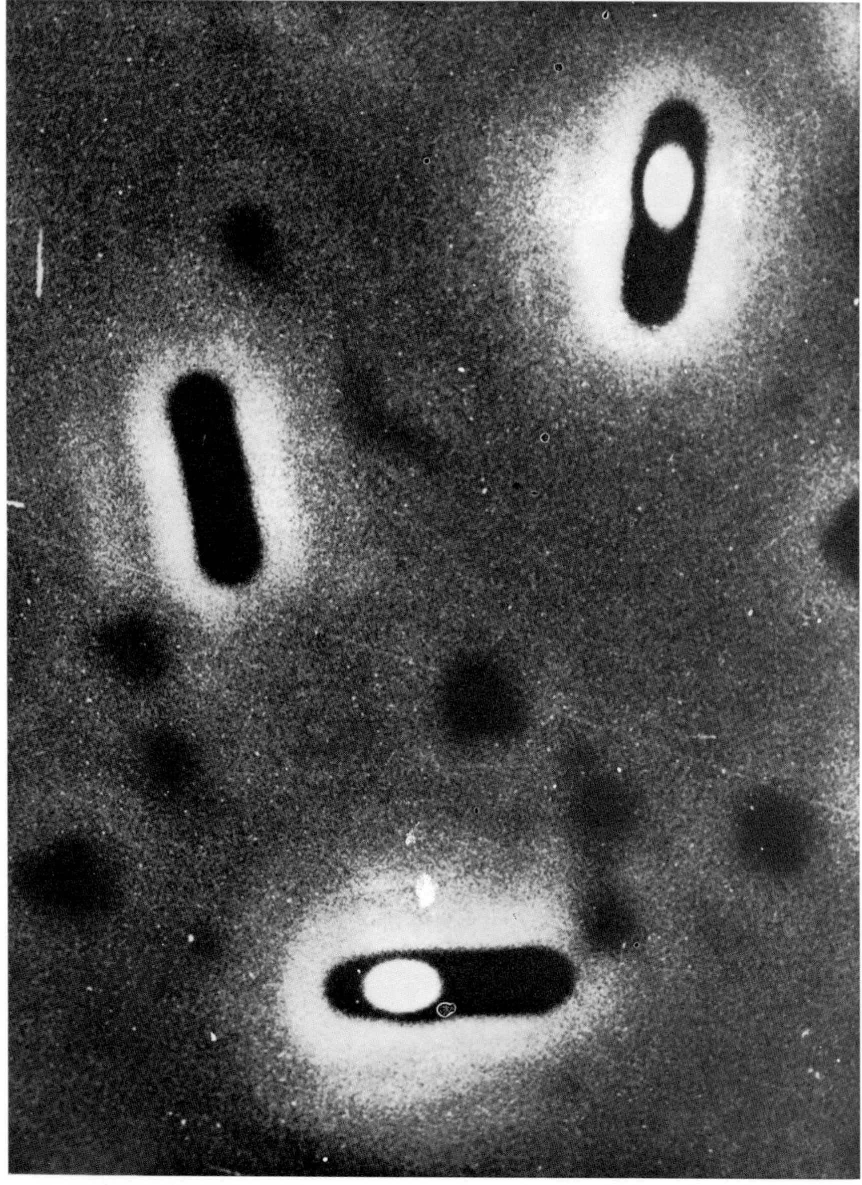

Figure 3.2
Spore and vegetative form of micro-organisms. A photomicrograph of *Clostridium botulinum* showing (left centre) an organism in its vegetative form, and two organisms (top and bottom) in the process of conversion into spores. The dark 'sausage' shaped bodies are the vegetative cells; the light oval area within the cell is the spore forming. (× 12500.)
Photo, Unilever Research Laboratory.

Growth of micro-organisms

Micro-organisms reproduce by a process of simple cell division. On reaching adult size the cell divides into daughter cells of approximately equal size. These grow to full size and the dividing process is repeated. Starting from one cell, in the second generation there are two cells, in the third generation four cells, in the fourth generation eight cells, and by the twenty-first generation we have over one million cells.

Many micro-organisms grow very rapidly and in favourable conditions each new cell may take only 20 minutes to grow to full size and divide. At this rate the original single organism can give rise to one million organisms in seven hours. This illustrates the speed at which bacteria can infest and spoil food if conditions are favourable to their growth.

Fortunately bacteria do not always grow at this great pace. A typical growth curve for bacteria is shown in figure 3.3 where the logarithm of the bacterial population is plotted against time.

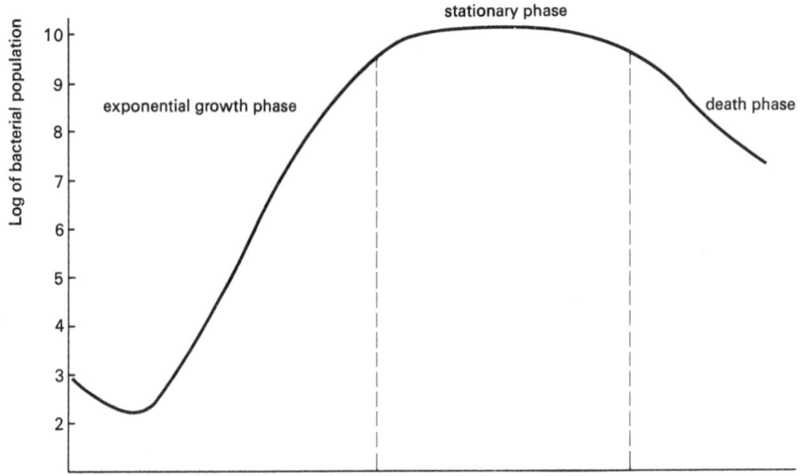

Figure 3.3
A typical bacterial growth curve.

At first, while the organisms adapt themselves to the medium, there is no increase in population, and often the count drops slightly. This period is known as the lag-phase. Then, there is a period of very rapid growth when the population doubles at regular intervals. This is known as the exponential phase or the logarithmic growth phase.

After some time the rate of growth falls away and finally stops. This slowing up is due to the nutrients being used up, or to the organisms becoming inhibited by the by-products of their metabolic processes. For example, many bacilli produce acid, until the pH of the medium drops to a point where no more growth occurs. This period is known as the stationary phase. In foods this is usually reached when the population is about 10^9 cells per cubic centimetre. Finally in the absence of fresh nutrients the population of live cells falls off. This is the death phase.

The rate of growth is controlled by a number of factors including the initial population, temperature, presence or absence of oxygen, moisture content, and chemical environment. With higher initial counts the lag-phase is shorter and fewer generations are needed in the exponential phase to reach a dangerous level, and the food spoils more quickly. This illustrates the importance of avoiding contamination as far as possible in the preparation of food.

Food spoilage bacteria can be divided roughly into three classes, according to the temperature at which they grow most quickly.

Thermophilic (heat-loving) bacteria grow most rapidly at temperatures above 40 to 45 °C and up to 75 °C. A number of spore-forming bacilli which cause souring are in this class.

Mesophilic bacteria are those which have their highest growth rate at temperatures between 25 and 40 °C. This class includes most organisms for which man and other warm blooded animals act as host, for example *Salmonella* and *Staphylococcus aureus*. Organisms of this class will grow below 25 °C but more slowly, and many plant-borne and soil-borne organisms are mesophiles.

Psychrophilic (cold-loving) bacteria are misnamed. They tolerate rather than love cold conditions. Many water-borne organisms are in this class. Their optimum growth rate may be at temperatures as high as 20 to 30 °C but they will continue to grow, although more slowly, at temperatures as low as 0 to 5 °C.

Within each of these classes there is a minimum temperature below which no growth takes place. Above this the rate of growth increases with temperature until the optimum temperature is reached. Above the optimum temperature, cell growth is inhibited and finally the cells are killed.

Bacteria can also be classified by their need for oxygen. Aerobic organisms will grow only in the presence of oxygen. Very many of the acid forming organisms such as *Lactobacilli* are in this class. Anaerobes such as *Clostridium botulinum* and *Clostridium welchii* will grow only in the absence of oxygen. Some organisms will grow either in the presence or absence of oxygen. These are known as facultative anaerobes, and include *Staphylococci*.

Bacteria require a moisture content of 20 to 40 per cent and will not grow in dry food. Moulds will grow at lower moisture contents. Bacteria require suitable nutrients and minerals, and most prefer a neutral or slightly alkaline medium. Acid tolerance varies considerably among organisms, but most can grow at pH values ranging from 4.5 to 10. Some yeasts and moulds will grow at a pH value as low as 1.5. Many organisms are inhibited by concentrations of 5 to 10 per cent of sodium chloride.

Non-microbial changes

Even without microbial damage foods undergo changes in storage: the surfaces of apple slices rapidly turn brown; sweet crisp lettuce soon becomes limp and bitter. These changes are due to biochemical reactions and occur even in the absence of micro-organisms.

In the previous chapter we discussed the cellular structure of food. The cells in biological tissues are, of course, not inanimate like the bubbles in foam rubber, but are highly organized and have very active biochemical systems. We can compare a cell in a living plant or animal to a highly organized and fully automated factory. The factory takes in fuel (oxygen, carbohydrates, and fats) and raw materials (water, minerals, amino acids, etc.), and converts fuel to energy which it uses to convert raw material to end products (proteins, polysaccharides, etc.). The manufacturing processes which it employs are normally multi-stage with a number of intermediate compounds. Finally the cell disposes of its waste products (carbon dioxide, ammonia). Harvest or slaughter has the effect on this factory of cutting off the supply of fuel and raw materials, either immediately or over a period of time, preventing the end products from being moved out of the factory, and blocking up the drains. If the factory is damaged, certain stages in the sequence of reactions will be stopped, and intermediates from previous reactions will accumulate.

This factory analogy must not be carried too far, but it may help you to realize the complexity and dynamic state of fresh foods. You cannot expect foods to behave simply as the bricks and mortar of a factory would behave. You must take into account the complex chemical reactions which are going on in the space enclosed by the bricks and mortar.

The time taken for food to spoil in the absence of micro-organisms is roughly inversely proportional to the activity of its biochemical systems at the time of slaughter or harvest. Wheat, for example, is harvested and stored at a stage in the life cycle of the plant when the activity of the cells is very low. The biochemical systems of wheat grains are designed to cope with the absence of nutrients, so that grain can exist without damage when removed from the parent plant. Other plant foods such as green peas are harvested at an 'unnatural' stage, when the cells are very active and not equipped to exist apart from the parent plant.

If we are to store food, we must find some way of preventing these undesirable changes occurring. Thus the end product of the agricultural industry becomes a raw material for another industry, that of food preservation.

As with changes produced by micro-organisms, not all biochemical changes in food are harmful. Meat becomes more tender and improves in flavour when it is hung after slaughter, and bananas picked green ripen to the usually preferred yellow form. However most biochemical and microbiological action leads to food spoilage.

Raw food has yet another shortcoming. Try to imagine what it would be like to live entirely on a diet of raw food! You might enjoy the fruit but how about raw meat or dry wheat grains. You would undoubtedly prefer roast meat and bread. In order to increase the palatability of food we deliberately change its form by cooking or processing. Once again agricultural end products form the raw material of another industry.

Cooking also makes some foods easier to digest, for example boiled milk is more easily digested than cold milk. It increases the safety of foods by destroying many harmful bacteria and toxic substances produced by bacteria. The desirable effects of cooking are accompanied by some undesirable reactions leading, for example, to losses of vitamins and flavour.

Summary

We can summarize the various types of food changes:

1 Due to microbial activity

a Undesirable, and sometimes dangerous to health

b Beneficial

2 Due to biochemical reactions
a Undesirable
b Beneficial

3 Due to cooking or food processing
a Beneficial
b Undesirable

The chemical and biological reactions which underly changes in food are discussed in the remainder of this chapter. These reactions are of much more than purely academic interest. The basic problems in food preservation and processing are to prevent undesirable changes occurring, and to bring about desirable changes. If we know the underlying reactions, our knowledge of chemistry will usually suggest ways of controlling them. For example, we may observe that a green vegetable is turning brown during cooking. This observation in itself does not suggest a method of preserving the green colour but, if we know that the fresh green colour is due to chlorophyll, and that chlorophyll on heating in acid solutions gives a brown product, and in alkaline solutions gives a green colour, we can immediately suggest raising the pH of the cooking water.

Knowledge of the underlying reactions in food is of great importance in food technology, particularly with the newer and more sophisticated processes. Often, during the development of a new product or process, or with a change in raw material supply, undesirable changes occur unexpectedly which spoil the product. If we understand what these changes are, in terms of chemical reactions, it is often easy to suggest a solution. If the underlying reactions are not known, it is often difficult and may require a trial and error approach.

A great number of traditional methods for preserving or processing food were of course discovered by chance, or by trial and error. Cooks with no knowledge of chemistry know that you can keep vegetables green by adding bicarbonate to the cooking water. The traditional methods have evolved over many thousands of years, but the need for new and improved methods of food preservation and processing is extremely urgent, and may be a matter of life or death for the populations of some of the developing countries. The problem of feeding the future world population is discussed in more detail in chapter 7. It is sufficient here to point out that a large quantity of potential food never reaches the people who need it, either because in its natural form it is unpalatable or poorly digested, or because it spoils before it can be eaten. The proportion of food which spoils may be as high as 10 per cent on average throughout the world. This is equivalent to a loss of 150 million tonnes of food every year, enough to feed the entire population of North America. To make matters worse, the proportion of spoilt food is highest where the need is greatest. In parts of West Africa, for example, probably half the food produced is wasted by spoilage.

With the population of the world increasing faster than food production, we cannot afford to wait for chance discoveries to add to the methods available for food preservation. We need rapid advances in food technology and, in order to achieve this, we need to understand the basic chemistry of food changes.

In the practical section of this book a relatively simple spoilage phenomenon, the browning of apple slices, is studied. This will give an opportunity to investigate the underlying chemical reactions, to suggest and try out possible ways of stopping the reactions, and consequently to preserve the colour of the slices.

2 Natural changes

This section discusses the reactions which occur in raw, unpreserved foods. Although, for convenience, bacteriological and biochemical changes are dealt with separately, the reactions involved are often very similar, whether they occur in the food itself or in the bacterial cells. It should be stressed that the individual reactions are almost invariably simple enzyme-catalysed reactions, often identical to those occurring in living tissues. However, while the reactions in living tissue are organized and the sequences of reactions are controlled, in food the cellular organization breaks down and the reactions become uncontrolled and do not follow the correct sequence. Although the individual reactions are simple, the overall changes in food may be complex and difficult to sort out, because of the great number of reactions which can occur simultaneously.

Nearly all these reactions are catalysed by enzymes, and we should recapitulate what we know about enzymes.

1 Enzymes are catalysts, and almost every reaction in the living tissue is catalysed by some enzyme.

2 Enzymes are highly specific catalysts; that is, they normally catalyse only one particular reaction.

3 Enzymes are proteins.

4 Enzymes are destroyed by heat.

5 Enzyme activity is dependent on pH and most enzymes are active only within a limited range of pH.

Undesirable microbial changes

The spoilage reactions brought about in milk by micro-organisms provide examples of many types of microbial reactions. Fresh milk, as it leaves the udder of a healthy cow, contains few bacteria and these do not grow well in milk.

As soon as it has left the udder it becomes contaminated from the exterior of the udder. Bacteria of soil and water may enter in this way. Other sources of contamination are the hands of the milker, the milk pails, milking machines, and dairy equipment. By the time bulk milk reaches a factory it often has a bacterial count of 10^7 organisms per cubic centimetre. Milk is an excellent culture medium for many kinds of micro-organisms because it is high in moisture, nearly neutral in pH, and rich in microbiological nutrients, such as milk sugar (lactose), butterfat, citrate, proteins, amino acids, and minerals.

Carbohydrate changes (souring)

A large proportion of the bacteria present in milk under normal conditions are lactic acid producing and the first major change which occurs is the conversion of lactose to lactic acid. This requires a large number of consecutive reactions. Lactose is first split to glucose and galactose by the enzyme lactase, which is produced by the lactobacilli. The monosaccharides are then converted by other enzymes to glucose-6-phosphate, which is converted into pyruvic acid by a sequence of ten reactions, involving eight enzymes. This reaction sequence is known as the Emden-Meyerhof-Parnass (EMP) glycolytic pathway. It plays a major part in metabolic processes in bacteria, in aerobic and anaerobic respiration in plants, and in animal muscle. It is also involved in a great many types of spoilage, both microbial and biochemical.

The final stage in milk souring is the hydrogenation of pyruvic acid to lactic acid, catalysed by the enzyme lactic dehydrogenase.

$$\underset{\text{pyruvic acid}}{CH_3COCO_2H} \xrightarrow[\text{lactic dehydrogenase}]{+2H} \underset{\text{lactic acid}}{CH_3CHOHCO_2H}$$

The pH of the milk drops as lactic acid forms and this affects the milk proteins. Fresh milk contains three main classes of soluble milk protein: casein, globulins, and albumins. Casein exists in a micellar form with calcium, inorganic phosphate, magnesium, and citrate. As the pH drops, the bonds with calcium and phosphate are broken and casein becomes less soluble, and at about pH 5.2 it coagulates and separates out. The phenomenon is known as curdling. Albumins and globulins remain in the liquid phase as serum or whey. After a time the decreasing pH inactivates the lactic acid producing bacteria. No further lactic acid is produced and the product left is known as curds and whey. Although milk souring is now regarded as undesirable in this country, this has not always been the case. Do you remember Little Miss Muffet? If the contaminating organisms in milk are predominantly lactic acid producers, sour milk is a very stable product and, in the past, was regarded as a useful form in which to store milk. It is still used quite widely in some countries.

If a high proportion of organisms other than lactobacilli are present, the spoilage does not stop at this point, and a lot of end products besides lactic acid may be produced, such as ethanol, acetaldehyde, acetic acid, acetone, carbon dioxide, and some higher alcohols, acids, and carbonyl compounds. For example, pasteurized milk, if stored at too high a temperature, spoils in a different manner from the souring of raw milk. The mild heat treatment of pasteurization kills off most of the normal lactic acid producing organisms, and although spoilage of pasteurized milk takes longer to develop, it is much more unpleasant than souring, as it involves protein breakdown and fat breakdown.

Protein changes

The first stage in protein spoilage is breakdown by bacterial enzymes to give a mixture of amino acids followed by further breakdown of the amino acids. One common mechanism is oxidative deamination (removal of NH_2 group) followed by decarboxylation (removal of CO_2).

$$\underset{\text{leucine}}{(CH_3)_2CHCH_2\underset{\displaystyle NH_2}{\underset{|}{C}}HCO_2H} \xrightarrow[\text{deaminase}]{+O} \underset{\text{4-methyl-2-oxopentanoic acid}}{(CH_3)_2CHCH_2\underset{\displaystyle O}{\underset{\|}{C}}CO_2H} + NH_3$$

$$\underset{\text{4-methyl-2-oxopentanoic acid}}{(CH_3)_2CHCH_2\underset{\displaystyle O}{\underset{\|}{C}}CO_2H} \xrightarrow[\text{decarboxylase}]{-CO_2} \underset{\text{3-methylbutanal}}{(CH_3)_2CHCH_2\underset{\displaystyle O}{\underset{\|}{C}}H}$$

$$\underset{\text{3-methylbutanal}}{(CH_3)_2CHCH_2\underset{\displaystyle O}{\underset{\|}{C}}H} \xrightarrow[\text{dehydrogenase}]{+2H} \underset{\text{3-methylbutan-1-ol}}{(CH_3)_2CHCH_2CH_2OH}$$

A similar oxidative reaction can release hydrogen sulphide from sulphur-containing amino acids. Direct decarboxylation gives amines. These reactions lead to gas production and very unpleasant smells of the kind described as 'putrid'.

Fat breakdown

Many micro-organisms contain enzymes which break down fats. The first stage in the breakdown is hydrolysis, catalysed by the class of enzymes known as lipases. The hydrolysis gives diglycerides, then monoglycerides, and finally glycerol, releasing free fatty acids at each stage.

In milk the fatty acids have chain lengths from C_4 (butyric acid) to C_{18} (stearic and oleic acids). The free fatty acids give rise to cheesy and soapy flavours.

The next stage is the oxidative splitting of the fatty acids, particularly unsaturated fatty acids, catalysed by lipoxidases.

$$\underset{\text{unsaturated fatty acid}}{R{-}CH{=}CH{-}R^1CO_2H} \xrightarrow[\text{lipoxidase}]{+O_2} \underset{\text{fatty acid peroxide}}{R{-}\underset{|}{C}H{-}\underset{|}{C}H{-}R^1CO_2H}\ (\text{O—O bridge}) \longrightarrow \underset{\text{long chain aldehydes}}{RCH{=}O + O{=}CHR^1CO_2H}$$

The aldehydes formed by oxidative splitting have chain lengths of C_4 to C_{14}. They have very intense odours and give rise to rancid flavours. Oxidation of fats occurs also in the absence of enzymes, but much more slowly.

The main forms of bacterial spoilage in milk may now be summarized.

Breakdown of carbohydrates (fermentation) follows a number of closely related pathways, giving rise to lactic acid, ethanol, acetic acid, carbon dioxide, acetone, and a number of higher acids and alcohols.

Breakdown of proteins (proteolysis) gives peptides and amino acids, followed by reaction of amino acids to give ammonia, hydrogen sulphide, carbon dioxide, amines, mercaptans, and aldehydes.

Breakdown of fats (lipolysis) gives fatty acids, which may be oxidized to long chain carbonyl compounds.

Although we have used milk as our example, bacteria spoil other foods such as fruits, vegetables, and meat by the same or very similar mechanisms.

Beneficial microbial changes

Cheese making provides a good example of the beneficial effects of bacteria.

Cheese manufacture

The first step is to start acid production in milk. On farms and in small dairies fresh milk can be used and the natural population of micro-organisms in the milk can be allowed to ferment the lactose. On a large industrial scale the bulk milk supplies contain a higher and more varied bacterial population. The milk is given a heat treatment to get rid of most of the micro-organisms and a culture of a desirable strain of lactic producing bacteria is added. After some acid has formed rennin is added. Rennin is a digestive enzyme obtained from the lining of the fourth stomach of calves. It reduces the solubility of casein so that the souring milk sets to a soft curd, similar to a junket. The slightly acid pH produced by the bacteria is necessary for rennin to act, and its optimum temperature is about 35 to 40 °C.

The curd is cut up and heated (scalding). This destroys most of the micro-organisms and stops the souring. It also causes the curd to separate from the whey. The curd is filtered off, pressed into blocks, and allowed to mature. During maturation the cheese becomes firm and the flavour develops. The flavour is characteristic of the type of cheese, the method of manufacture, the type of organism present in the original milk, and the organisms such as moulds which inhabit the storage rooms. Different microbial populations create the very different but extremely characteristic flavours of the various types of cheese, such as Cheddar, Stilton, and Camembert.

Although the reactions which produce flavour, and indeed the actual compounds responsible for the flavours, have not been fully elucidated, the general reactions proceed according to the following scheme.

Proteins ⟶ peptides ⟶ amino acids ⟶ amines

Fats ⟶ fatty acids ⟶ ketones
aldehydes
alcohols
esters

Proteolysis and lipolysis are essential for cheese production but are regarded as undesirable in most foods.

Undesirable biochemical changes

Biochemical changes in food, both beneficial and undesirable, go on in the complete absence of bacteria. The chemical pathways involved are often identical or very similar to those in micro-organisms.

One undesirable change is pectin breakdown in fruit and vegetables which makes them mushy. Pectins play an important role in giving structure and texture to fruit and vegetables. In developing fruit and vegetables, galacturonic acid polymers are formed as long chains of polygalacturonic molecules to give an insoluble compound, protopectin. As fruit ripens, an enzyme, protopectinase, becomes active and hydrolyses protopectin to pectin.

Another enzyme, pectin methyl esterase, progressively hydrolyses the methoxy groups in pectin to give low-methoxy pectins and finally polygalacturonic acid. Other enzymes then catalyse the hydrolysis of the polygalacturonic acid chains to single D-galacturonic acid units.

During this sequence of enzymic reactions the pectic substances become more and more soluble, lose their gelling properties and become less and less capable of maintaining the rigid structure of the tissues. In a fruit, such as apple, this corresponds to the ripening process, by which an apple starts off by being too hard for eating, reaches a desirable texture for eating, then becomes over-ripe, over-soft, and finally the tissues collapse. All these processes are part of the life-cycle of the plant and are essential for the propagation of the species. The enzymes become active in a particular order and at a particular stage of development. However, when the plant is used as food, all the changes beyond the best condition for eating are highly undesirable and must be prevented if the food is to be stored.

The same series of reactions can be achieved, often more quickly, by pectic enzymes in some micro-organisms such as the moulds and rotting fungi found on damaged fruit.

Carbohydrate breakdown can be another undesirable change. Normally, in growing plants the energy required for cell building is obtained from the oxidation of saccharides produced by photosynthesis. The breakdown follows the EMP pathway we have already discussed to give pyruvic acid. Under normal aerobic conditions pyruvate is oxidized indirectly to carbon dioxide and water by a complicated series of reactions which enables the plant to harness the energy released in the oxidation. When a plant is bruised, for example in mechanical podding of peas, many cells are cut off from oxygen or are so badly damaged that the sequence of reactions required for aerobic respiration is completely disorganized. Under these conditions pyruvate is broken down anaerobically by reactions identical or similar to the bacterial processes described previously. The tissues accumulate ethanol, acetaldehyde, lactic acid, and other compounds with undesirable flavours.

Beneficial biochemical changes

The maturation of meat is an example of a desirable change. Maturation is the sequence of reactions which convert living muscle into hung meat. Freshly killed meat tends to be tough and has very little flavour. In normal practice carcasses are hung in cool stores for a few days, until the meat becomes tender and flavour develops. This improvement in quality is due to alterations in the system of enzymic reactions when no more oxygen is supplied to the muscle.

In life, the energy required to contract muscle (i.e. to do work) is available in adenosine triphosphate (ATP) which is a complicated molecule consisting of adenine (a purine), ribose (a pentose), and three molecules of phosphoric acid.

ATP takes part in a very great number of biological phosphorylations, giving up one molecule of phosphoric acid to give ADP (adenosine diphosphate). In other reactions (dephosphorylation) ADP takes up a molecule of phosphoric acid from an organic phosphate and reforms ATP. This is an essential mechanism for energy transfer in biological systems.

Muscular contraction is accompanied by dephosphorylation of ATP to ADP. Before any more work can be done the ADP must be reconverted to ATP. This is done by a multi-stage oxidation of glycogen, the storage carbohydrate of muscle. This oxidation system is very much the same as that of aerobic respiration in plants, consisting of hydrolysis of glycogen to glucose, the EMP pathway, and the aerobic oxidation of pyruvic acid. The energy made available by the oxidation of pyruvic acid is harnessed by the reconversion of ADP to ATP; oxidation of one molecule of pyruvic acid can give as many as 15 molecules of ATP.

When oxygen is cut off by death the available ATP is converted to ADP and the muscle stiffens (and may contract), giving the condition known as rigor mortis. The breakdown of glycogen can occur in anaerobic conditions as far as pyruvic acid, but oxidation of pyruvic acid is no longer possible. Thus there is no mechanism for reconversion of ADP to ATP and the muscle remains in its inelastic state. The pyruvate which accumulates is converted by muscle lactic dehydrogenase to lactic acid.

The build-up of lactic acid lowers the pH in the muscle. It normally drops to 5.4 to 5.5, where the glycolytic enzymes cease to be active. If animals have been starved, exercised, or frightened before slaughter, the glycogen reserves in the muscle will be low and, after death, all the glycogen will be used up before the pH drops to 5.5. This results in meat with different, and generally less desirable, properties.

After glycolysis stops, slower processes of maturation take over. At about pH 5.5, proteolytic enzymes are released from previously inactive combination with other compounds, and become active. The acid pH also tends to make the proteins more susceptible to proteolysis, which starts to break down the muscle structure and increases the tenderness.

Other breakdown products contribute to the flavour of cooked meat, particularly the breakdown products of nucleic acid and nucleotides (such as ribose from ATP), and amino acids.

3 Changes induced by processing

Cooking and food processing are mainly concerned with making food easier to digest, preserving it, and improving the flavour and texture. Cooking can be very simple, such as boiling milk to denature and coagulate the milk proteins and make them more easily attacked by the digestive proteolytic enzymes. It can be a great art, exemplified by the blending of many ingredients by skilled chefs to give exquisite flavours to their dishes. Or it can be a gigantic industrial process such as baking bread.

Beneficial changes

Bread making

As an example of deliberate control of chemical reactions in food we will consider the reactions involved in baking. The starting point of baking is dough, a mixture of flour and water. Wheat flour contains amylases which are capable of hydrolysing the amylose and amylopectin of starch. The hydrolysis does not occur to any significant extent in dry flour but begins immediately a dough is made. The most important product of the hydrolysis is maltose, a disaccharide.

When a dough is heated, the physical form of the starch is greatly changed. The starch in flour is contained in granules which form a suspension in water but do not dissolve. The starch granules take up 25 to 30 per cent of water but this has little effect on the structure of the granule, and is reversible if the granules are dried again. On heating to a temperature which will vary according to the type of starch, the granules suddenly swell and take up a large amount of water. The granules change in appearance, and soluble starch molecules begin to leak out of the granules. On further heating the suspension becomes more translucent and more viscous, the granule becomes enormous and ruptures, releasing more free starch. The viscosity decreases on further cooking but increases again on cooling. Gelatinized starch is much more readily hydrolysed by the starch splitting enzymes in the human digestive system.

A series of experiments illustrating the chemical and physical changes in starch during baking are given in the experimental section of this book.

A type of bread, known as unleavened bread, can be made by baking the simple flour–water mixture. This is the forerunner of modern bread, and lacks the honeycomb structure which gives modern bread its lightness. This structure is obtained by generating a multitude of small pockets of carbon dioxide throughout its bulk. The carbon dioxide is produced before the bread is baked and while it is still an elastic dough.

The oldest and most important method of aeration of the dough is fermentation. Yeast is added to the dough and converts the sugars naturally present in the flour, and the maltose made available by the action of amylases, into glucose and, by the fermentation systems we have already discussed, into alcohol and carbon dioxide. The carbon dioxide aerates the dough while the alcohol is driven off during baking.

Two of the proteins contained in flour, gliadin and glutenin, form an elastic complex called gluten when the flour is kneaded with water. Gluten forms an interconnected network which contains the carbon dioxide within the loaf. When the bread is baked the carbon dioxide expands, the starch gelatinizes, and the gluten coagulates to produce a more or less rigid loaf.

Salt is also added to the dough for making bread. It influences the rate at which fermentation takes place and enables the baker to control the development of the dough. In addition, it has a strengthening and toughening action on the gluten, possibly due to its inhibiting action on protein-splitting enzymes which, in the absence of salt, would cause a certain amount of degradation of the gluten.

All the ingredients are mixed thoroughly until a homogeneous dough is obtained. This is allowed to ferment for about one hour at about 25 °C. The dough is then thoroughly kneaded to expel some of the carbon dioxide and to bring the yeast cells into contact with more nutrient. It is then allowed to ferment for a further period during which the kneading process may be repeated. After fermentation the dough is divided into loaves. Much of the gas is expelled during this moulding process, and after being placed in baking tins or on baking sheets the dough is allowed a further short period of fermentation so that it may once again become inflated with carbon dioxide.

Bread is baked at 120 °C for 30 to 50 minutes depending on the size of the loaf. During baking the dough first expands rapidly because the pockets of carbon dioxide in the loaf expand as temperature increases. At first there may also be some slight increase in the activity of the yeast resulting in increased gas production, but this diminishes as the temperature increases, until at a temperature of about 55 °C the yeast is killed and fermentation ceases. As the temperature increases, the water present causes the starch grains to swell and gelatinize, and during this period the starch probably extracts some water from the gluten. Hot gluten is soft and devoid of its characteristic elasticity, and gelatinized starch now supports the structure of the loaf. The gluten begins to coagulate at about 75 °C and the coagulation continues slowly to the end of the baking period. The temperature of the interior of the loaf never exceeds the boiling point of water, in spite of the high temperature of the oven. Water and much of the carbon dioxide and alcohol formed during fermentation escape during baking.

Browning reactions

One change which accompanies bread making occurs very widely in cooked foods and this is the brown colour which forms on the outside of a loaf. In bread making and pastry making this colour change is pleasant so long as it only occurs on the outside of the bread or pastry. In other foods such as cooked green vegetables, the browning reaction is highly undesirable.

The browning which takes place in cooking or baking is quite different from the browning which develops in some raw fruit. Only the first stages in the reactions have been elucidated with any certainty. The most important reaction leading to the formation of browning pigments is known as the Maillard reaction. Aldehydes, ketones or reducing sugars combine readily with amino acids, peptides, and proteins, to form first a Schiff's base, and then an N-substituted glycosylamine.

$$\begin{array}{c} \mathrm{HCO} \\ | \\ (\mathrm{HCOH})_n \\ | \\ \mathrm{CH_2OH} \end{array} + \mathrm{RNH_2} \rightleftharpoons \begin{array}{c} \mathrm{RNH} \\ | \\ \mathrm{HCOH} \\ | \\ (\mathrm{HCOH})_n \\ | \\ \mathrm{CH_2OH} \end{array} \overset{-\mathrm{H_2O}}{\rightleftharpoons} \begin{array}{c} \mathrm{RN} \\ \| \\ \mathrm{CH} \\ | \\ (\mathrm{HCOH})_n \\ | \\ \mathrm{CH_2OH} \end{array} \rightleftharpoons \begin{array}{c} \mathrm{RNH} \\ | \\ \mathrm{HC}- \\ | \\ (\mathrm{HCOH})_{n-1}\,\mathrm{O} \\ | \\ \mathrm{HC}- \\ | \\ \mathrm{CH_2OH} \end{array}$$

sugar — amino compound — addition product — Schiff's base — N-substituted glycosylamine

Obviously, only carbohydrates which have a free carbonyl group can take part in this reaction.

The next step is a rearrangement of the molecule into a non-cyclic ketose.

After these first two stages the products are still colourless. The next stage begins with the loss of carbon dioxide from the system, apparently from the amino acid (R) part of the molecule, and discoloration starts. The reactions leading to the brown pigment are still not clear.

The sequence of reactions can be blocked by adding sulphite to the original mixture. The sulphite combines with the free carbonyl groups and blocks any formation of a Schiff's base.

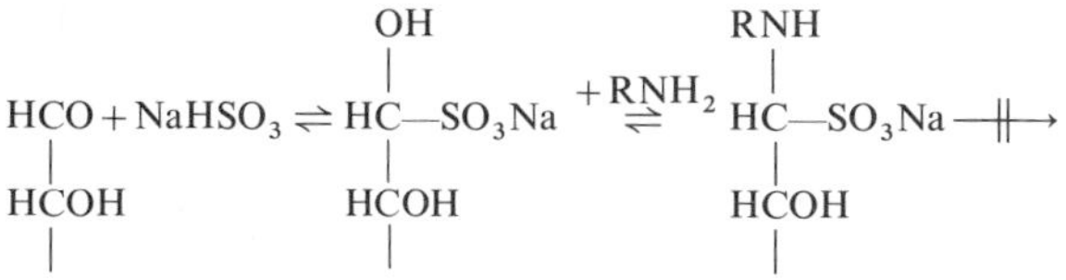

```
                       OH                      RNH
                       |                        |
HCO + NaHSO3  ⇌   HC—SO3Na    +RNH2 ⇌   HC—SO3Na —||→
 |                     |                        |
HCOH                 HCOH                     HCOH
 |                     |                        |
```

Sulphites are widely used in the food industry to prevent undesirable browning.

Meat colour

The principal pigment in muscle cells is myoglobin, a protein closely related to the haemoglobin of red blood cells. It is a combination of a protein (globin) with haem, an iron-containing porphyrin-ring structure.

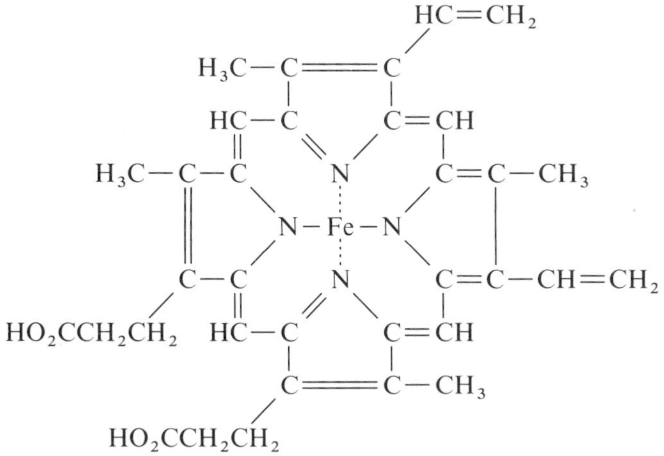

```
                                  HC=CH2
                                 /
               H3C—C=====C
                   |     |
               HC—C      C=CH
               ||   \\  /    |
       H3C—C—C      N      C=C—CH3
           ||    \   :   /     |
           ||     N—Fe—N       |
           ||    /   :   \     |
            C—C      N      C=C—CH=CH2
           /  ||   //  \    |
HO2CCH2CH2   HC—C      C=CH
                 |     |
                 C=====C—CH3
                /
       HO2CCH2CH2
```

Figure 3.4
The structure of haem.

In myoglobin the iron in the protoporphyrin ring is in the iron(II) oxidation state. The unique property of haemoglobin and myoglobin is their capacity to bind a molecule of oxygen without a change in the oxidation state of the iron. The reaction is readily reversible and is most important in the transfer of oxygen to the tissues in the living animal. In fresh meat the tissues away from the surface contain myoglobin which is purplish-red. At the surface of freshly cut meat, oxygenation gives oxymyoglobin which is bright red.

In addition to the oxygenation reaction the iron(II) in myoglobin can be oxidized to iron(III). This gives metmyoglobin which is brownish. These reactions are shown in abbreviated formulae.

O_2 N N Fe^{2+} N N globin	$-O_2$ $\rightleftharpoons$ $+O_2$	H_2O N N Fe^{2+} N N globin	oxidation $\rightleftharpoons$ reduction	N N Fe^{3+} N N globin
oxymyoglobin (bright red)		myoglobin (purple-red)		metmyoglobin (brownish red)

The formation of metmyoglobin leads to brownish discolorations in ageing meat.

On cooking, the protein becomes denatured and separated from the protoporphyrin ring which is converted into brownish pigments, giving the colour to cooked meat.

Myoglobin reactions are also responsible for the colour of bacon. In the traditional bacon curing process, nitrate is added to the curing brine. Bacteria in the brine reduce nitrate to nitrite, which reacts with myoglobin to give nitrosomyoglobin which is the bright red colour of bacon.

Undesirable changes

Discoloration of vegetables

The fresh bright green colour of vegetables is due to chlorophyll. As vegetables are cooked the brightness of the colour fades, becomes dull olive-green, then yellow-green and finally becomes brownish. The rate and extent of this reaction depends on the pH of the cooking water.

The structure of chlorophyll is similar to that of haem, the main differences being that the central metal in the porphyrin ring is magnesium instead of iron, and that one of the terminal —$(CH_2CH_2CO_2H)$ groups is esterified with phytol ($C_{20}H_{39}OH$), an unsaturated alcohol.

On warming chlorophyll in slightly acid solution, the magnesium atom is removed from the porphyrin ring to give a brownish-green compound. On further heating or in stronger acid solution the phytol residue is hydrolysed off the rest of the molecule to leave a brown compound.

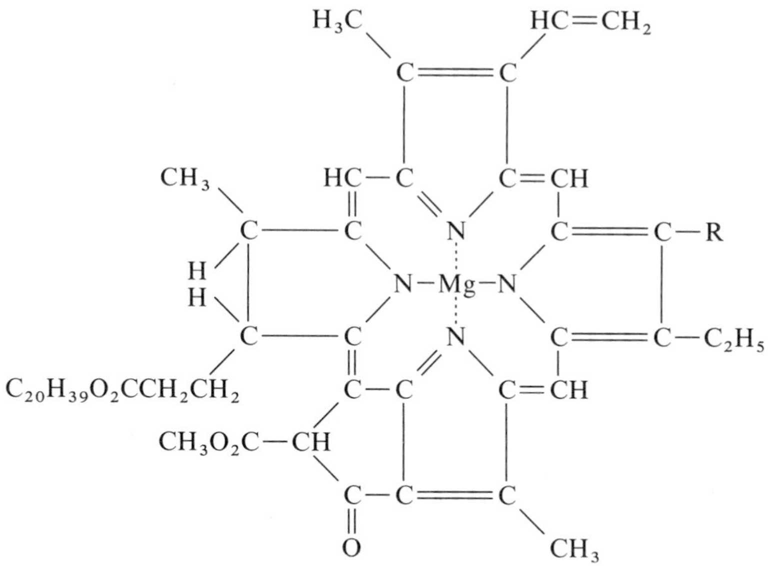

H_3C $HC{=}CH_2$
C=C
CH_3 HC—C C=CH
C—C N C=C—R
H
H N—Mg—N
C—C N C=C—C_2H_5
$C_{20}H_{39}O_2CCH_2CH_2$ C—C C=CH
CH_3O_2C—CH
C—C=C
O CH_3

Figure 3.5
The structure of chlorophyll. In chlorophyll (a) R is —CH_3; in chlorophyll (b) R is —CHO.

On heating chlorophyll in alkaline solution, the phytol residue is removed without removing the magnesium atom, to give an olive-green compound. The removal of phytol can also occur in vegetables during storage by the action of the enzyme chlorophyllase. Because alkaline conditions help to preserve the green colour, sodium bicarbonate is often added to the cooking water for vegetables.

Changes in pectins

When vegetables are heated in water, the pectins become more soluble and some are extracted into the water. The effects on the intercellular pectins, which act as the cement in vegetable structure, are the most important. The more soluble the pectins become the softer and more floury the texture. This softening is desirable to some extent, but even slight overcooking gives unpleasantly mushy vegetables.

When the pectins are hydrolysed by cooking to low-methoxy pectins, they can be precipitated by calcium and magnesium ions. These ions are present in hard water, and for this reason vegetables cooked in hard water are firmer or tougher than those cooked in soft water. Alkaline conditions tend to soften vegetables by increasing hydrolysis of pectins and by removing calcium. Thus the bicarbonate added to preserve the colour of green vegetables also softens them and if too much is added the effect can be disastrous, particularly in soft water.

Changes in proteins

We have mentioned earlier that denaturing proteins by heat tends to make them more digestible. An example of this denaturation is the setting of egg-white in a frying pan. The increase in digestibility however applies only to mild heat, and prolonged heating makes protein hard and indigestible. As the proteins are denatured they lose much of their capacity to bind water and this results in loss of liquids from cooked foods. Cooked meat, even stewed meat, contains less water and hence less of its soluble ingredients than fresh meat.

Losses in flavour and nutritive value

As we have mentioned in the sections on pectins and proteins, cooking foods leads to losses of water and soluble material from the food into the cooking water. Volatile flavours are driven off by heating and much of the fresh flavour is lost. This is of little consequence in meat, where most of the desired flavour only appears on cooking, but is generally undesirable in fruit where, for example, the finest flavours of fresh strawberries and raspberries are lost on cooking. However, many people prefer cooked (canned or frozen) fruit which often has a distinct flavour of its own. The choice between a cooked and fresh fruit flavour is a matter of opinion.

Several vitamins, particularly in the B group and vitamin C, are destroyed by heat, by oxidation, or by leaching into the cooking water. Fresh peas cooked for ten minutes in boiling water lose 20 to 40 per cent of their original vitamin C content, depending on variety and maturity. Major soluble ingredients are also lost to the cooking water. For example, 25 to 50 per cent of the sugar in young green peas is lost by boiling for a few minutes. On the industrial scale the loss of weight in cooking is extremely important. To produce 1 kg of product the manufacturers have to buy considerably more than 1 kg of raw material, a point not always remembered by consumers when comparing prices of fresh and processed foods.

Chapter 4

Principles of food processing and preservation

The need to preserve and process foods has been discussed in the preceding chapters. We have seen that raw foods are not inert, and unless we take some positive action they will deteriorate through the action of micro-organisms or of the native enzymes in the food. The processes of spoilage take place by a series of reactions, the great majority of which are catalysed by enzymes in the food or in micro-organisms. A large proportion of the reactions, particularly in the earlier stages, are hydrolytic or oxidative.

Food preservation, to be successful, must either stop these reactions or greatly reduce the rate at which they occur. Now that we know the reactions involved we can draw on our knowledge of the physical chemistry of reactions, and suggest ways in which these adverse changes can be reduced. The rate of a reaction can be reduced by:

1 Removing or excluding a reactant
2 Removing or inactivating the catalyst (or micro-organism)
3 Lowering the temperature
4 Altering the reaction system

How can these aims be achieved practically?

1 In order to remove or exclude a reactant we can
a exclude or remove oxygen
b exclude or remove water (to prevent hydrolytic reactions)
c adjust pH to reduce H^+ or OH^- concentration

2 In order to remove or inactivate a catalyst or micro-organisms we can
a heat
b add an enzyme inhibitor or microbial inhibitor
c adjust pH to a region where enzymes or micro-organisms are inactive
d prevent micro-organisms from entering the system (i.e. establish aseptic conditions)

3 In order to lower the temperature we can
a cool
b freeze

4 In order to alter the reaction system we can

a remove solvent and change to the solid phase, i.e. dry or freeze

b change the ionic concentration of the solution, for example by adding salt

c add a competitive reactant, for example an antioxidant which will compete with the substrate for any oxygen in or entering the system

In processes such as cooking or baking, we are primarily concerned with promoting wanted reactions, rather than stopping undesirable reactions. We increase the rates of certain reactions, usually by heating.

The investigations in the practical section of this book give an opportunity of identifying the important reactions in a particular form of food spoilage, and of suggesting and trying out methods of controlling these reactions.

Obviously food technology is not such a simple matter as merely identifying and controlling single reactions. For example, if an undesirable reaction is enzymic, practically every method of reducing the rate of reaction will reduce the rate of all other enzymic reactions in the food, some of which may be desirable.

Some of the methods of controlling reactions affect the reaction system in more than one way. Freezing, for example, lowers the temperature, removes water, and converts the system to the solid phase; all these effects act in the same direction. On the other hand the first effect of heat is to increase the rate of reaction, although it is rapidly counteracted in biological reactions by the thermal destruction of enzymes or micro-organisms. Cooking (i.e. heating) is primarily intended to increase palatability, but it also gives some degree of preservation.

Anything one does to food normally affects more than one of the factors which go to make up quality, that is, colour, flavour, texture, appearance, palatability, and wholesomeness. It is very difficult, for example, to alter the texture without altering the colour and flavour at the same time. In addition, a process is seldom wholely beneficial; for example, canning overcooks, irradiation sterilization produces 'off' flavour, and attempts to improve the colour of green vegetables by pH control also causes undesirable texture changes. One has to improve *quality* rather than a single factor.

The skill in food processing lies in getting the best quality by adjusting colour, flavour, and texture, but without harming the consumer by reducing the nutritive value or introducing toxic material, and without making the food too expensive.

We can now recognize industrial food processing and preservation techniques as being applications of the simple principles of controlling rates of reaction.

Cooking. Destroys enzymes and most micro-organisms.
Gas or vacuum packing. Removes or excludes a reactant (oxygen), and inhibits aerobic micro-organisms.
Canning or bottling. Destroys enzymes and micro-organisms, and removes and excludes reactants (oxygen).
Chilling (domestic refrigeration). Slows reactions.
Freezing. Slows reactions, removes a reactant (water), and converts the system to the solid phase.
Dehydration. Removes a reactant (water), and converts the system to the solid phase.
Curing. Alters ionic concentration of the solution, and inhibits enzymes and micro-organisms.
Pickling. Alters the ionic concentration of the solution, reduces the pH, and inactivates enzymes and bacteria.
Irradiation. Destroys micro-organisms.

We can now discuss the main features of these processes, as they affect the spoilage reactions.

Cooking

We can look on cooking as the simplest form of heat treatment for foods. It often dries the food to some extent, but is not continued until the food is dehydrated. Cooked food is not normally covered in a way that would exclude oxygen or micro-organisms.

Cooking is a three stage process: the food is first warmed up, then held at the maximum cooking temperature for some time, and finally cooled. In order to minimize enzymic and microbiological damage it is important to keep the warming and cooling times as short as possible. As food warms up each enzymic reaction in turn becomes more rapid until the optimum temperature for the enzyme is reached. Above the optimum temperature the rate falls off rapidly, and stops when the enzyme is denatured. However, if the heating up is slow, quite a lot of enzyme action may occur before the enzymes are destroyed. Pectic enzymes are particularly active and the rate of initial heating may affect the consistency of a plant product quite significantly. Similarly, some micro-organisms, particularly the thermophilic bacteria, become more active as the temperature increases and then decline in activity as they are killed by the heat.

Most enzymes are destroyed quickly above 70 to 80 °C and five minutes at 100 °C will certainly destroy all enzymes. It will also kill most micro-organisms, but some spore-forming bacteria will survive very much longer periods of heating. For example, spores of the dangerous pathogen *Clostridium botulinum* may survive five hours at 100 °C. The time required to kill bacteria varies very considerably with pH. Figure 4.1 shows the heat resistance at 100 °C of spores of *Clostridium botulinum* in buffers of varying pH.

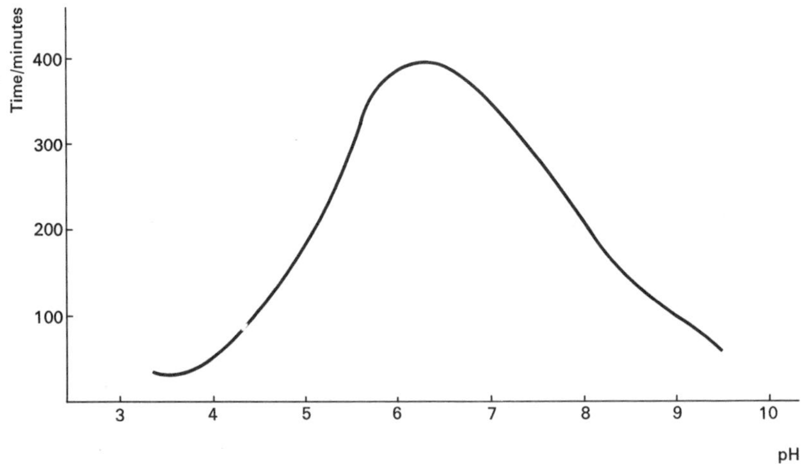

Figure 4.1
Heat resistance of spores of *Clostridium botulinum* at 100 °C, in buffers of varying pH.

The heat resistance drops off very rapidly below pH 5, and below pH 4.5 the bacteria are relatively easily destroyed. For this reason high acid foods such as fruit (pH 2.5–4.3) require less cooking than vegetables (pH 5.2–6.0), meat and poultry (pH 5.5–6.2), and particularly sea foods (pH 6.4–7.0), to make certain of killing off the bacteria.

The rate of heat penetration into the interior of pieces of food is an important consideration in determining the time needed to destroy enzymes or bacteria, particularly in food cooked in large pieces such as loaves of bread and joints of meat. Most food has a low thermal conductivity. For example, beef has a thermal conductivity of 268×10^{-5} joules s^{-1} cm^{-1} $°C^{-1}$ very similar to wood (200×10^{-5}) and asbestos (148×10^{-5}). Sponge cake has a thermal conductivity of 80×10^{-5}. In fact foods, particularly baked foods, are good insulators. The low conductivity is illustrated when toasting bread, when one side of the slice can be on fire without discolouring the other side.

This results in the centre of the food being often at a much lower temperature than the outside of the food. Even when food is immersed in boiling water it still takes a long time for the centre of the food to reach 100 °C. In ovens, although the temperature of the air may be about 200 °C, the temperature at the centre of the portion of food will be lower than 100 °C unless the cooking is prolonged. The temperature can be raised by pressure cooking. At a pressure of 0.10 N mm^{-2} above atmospheric the boiling point of water is 120 °C and the temperature in the interior of the food will be correspondingly higher than in normal boiling or steaming. Cooking in steam rather than in air also increases the rate of temperature rise within the food, because heat transfer from steam is much more effective than heat transfer from air, and also steam prevents the formation of a dry surface skin with a very low thermal conductivity.

The low thermal conductivity of food is an important factor in the last stage of cooking, the cooling stage. If the food is to be eaten hot, slow cooling is an advantage. However, if the food is intended to be eaten cold after a period of storage, the slow cooling can be a hazard.

Although the food may be free from bacteria immediately after cooking it can rapidly become recontaminated with bacteria from the cook's hands, from plates or dishes, from other food in an overcrowded larder, and from airborne particles. Unless food is cooled rapidly and kept cool there will be a long period when the surface is at an ideal temperature for bacterial growth. This is a common cause of food poisoning.

Gas or vacuum packing

In the most common form of gas or vacuum packing the food is put into a pouch or envelope made of plastic or aluminium foil. The pouch is swept out with nitrogen or evacuated and it is then sealed so that it is airtight. This process obviously does nothing to prevent the activity of the majority of enzymes. It does prevent or greatly reduce non-enzymic oxidations, and is helpful in storing foods with a high fat content, such as potato crisps. It prevents the growth of aerobic bacteria, but encourages the growth of anaerobic bacteria. This can lead to a potentially dangerous situation unless attention is paid to hygiene in preparing and packaging the food. Under normal conditions of aerobic storage and with normal bacteriological contamination, the harmless bacteria which cause spoilage ('off' odours) outgrow the dangerous pathogens, so that it is almost certain that a consumer would reject the food because it looked or smelt 'off', long before the pathogens had reached dangerous levels. Thus many spoilage organisms provide a natural early warning system. However, under anaerobic conditions the anaerobic pathogens may outgrow the normal spoilage organisms, and the pathogens reach a dangerous level without any visible signs of spoilage. For this reason, gas or vacuum packing is reserved for foods which are, in themselves, poor media for bacterial growth, such as dried food (potato flake) or food with a high salt content (bacon).

Canning

Canning combines the principles of cooking and vacuum packing, by destroying all enzymes and bacteria and sealing the container so that no air can enter and no recontamination with bacteria can occur. Cans are filled with food and they are heated in steam or in a boiling water bath so as to replace the air above the food with steam. The lid is then sealed and the cans are heated in autoclaves under pressure. After heating for the required time the cans are cooled as rapidly as possible.

The heating time in commercial canning has to be controlled far more carefully than in ordinary cooking. A cook normally is not consciously trying to kill bacteria; food is cooked to make it taste as good as possible. In commercial canning, on the other hand, the minimum heating time is determined by the heat required to ensure that there can be no surviving bacteria. Unfortunately some foods, particularly low acid foods, require longer to achieve sterility than the optimum time for palatability and many canned foods are overcooked. In some cases the overcooking is more than is acceptable and some vegetables cannot be canned successfully.

The heating time required to destroy bacteria in commercial canning depends on a number of factors: the cooking temperature; the pH of the food; the salt and sugar content; the level of initial bacterial contamination; the size of the can; the thermal conductivity; and the physical form of the food. The bacteria which concern us most are the relatively heat resistant spore-formers such as *Clostridium botulinum.*

The time required to kill bacteria decreases rapidly with increasing temperature. The table below gives the time required to kill spores of *Clostridium botulinum* in pH 7 buffer at various temperatures.

Time /minutes	Temperature /°C	Pressure above atmospheric/10^{-2}N mm^{-2}
330	100	0.0
100	105	2.1
33	110	4.1
10	115	6.8
4	120	9.6

In commercial practice the correct heating time is determined by adding a culture of a particularly resistant organism to cans of the product being tested. An organism which is more heat resistant than *Clostridium botulinum* is chosen. Cans are heated for various lengths of time at a certain temperature (T) and the number of surviving bacteria per gramme determined. If the log of the number of surviving bacteria is plotted against time, it gives a straight line, as shown in figure 4.2.

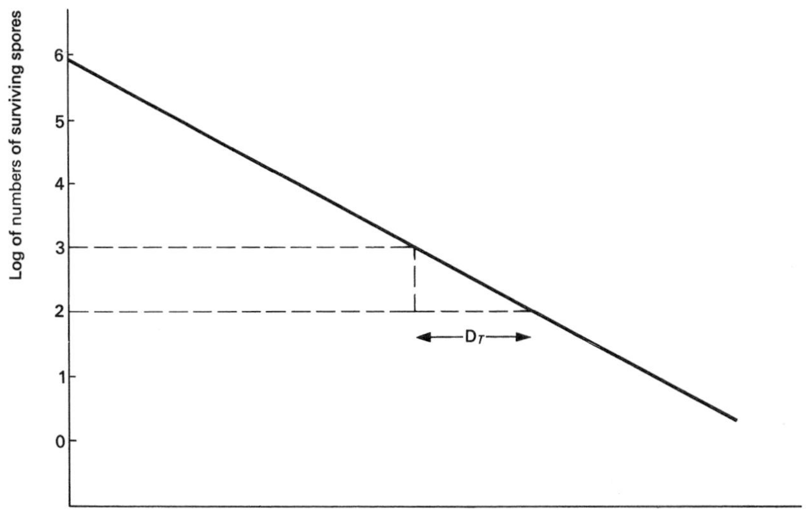

Figure 4.2
Spore survival rates after heating for various times at temperature T.

This graph gives the time needed to reduce the bacterial population ten-fold at temperature T. This time is known as D_T, the 'Decimal reduction time' at temperature T. The higher the temperature T, the lower is D_T.

Safe heating times at temperature T can be calculated once D_T and the likely initial population of bacteria are known. If the original population was, say, 10^4 organisms per gramme and $D_T = 1$ minute, then, after heating for 1 minute at T °C there will be 10^3 surviving organisms per gramme. After heating for 4 minutes ($4D_T$) there will be 1 organism per gramme.

These bacterial counts can be expressed in a more useful way by considering *not* the number of organisms per gramme, but the number of grammes of food per organism. Thus in the example given above the original distribution of bacteria would be 1 organism in 10^{-4} grammes. After 1 minute of heating there would be 1 organism in 10^{-3} grammes and after 4 minutes 1 organism per gramme.

If the heating is continued for a further 3 minutes, there would then be 1 organism per 1000 g of food. If one can contains 100 g of product this means that of 10 cans examined after 7 minutes heating, one can will contain 1 organism and the other nine cans will be completely free of surviving bacteria. We can then say that after 7 minutes ($7D_T$) at T °C the chances of finding a surviving organism in a can is 1 in 10; after 8 minutes 1 in 100; and after 9 minutes 1 in 1000.

The odds on finding a surviving organism in a can obviously depend on the initial population. Food handled correctly before canning is most unlikely to have a population of *Clostridium botulinum* greater than 10 to 100 organisms per gramme, and in a 500 g can there might be a maximum of 5000 organisms. Normal commercial practice is to give a heat treatment of $12D_T$ minutes at temperature T. Even with the maximum probable initial population this heat treatment reduces the probability of a can containing a surviving spore to 200 million to 1 against. As the majority of food has much less than 100 organisms per gramme to start with, the odds against being poisoned by *Clostridium botulinum* in canned food is something like 10000 million to 1, which has been said to be less than the chance of being hit by a can of food dropped from an aircraft. These figures, of course, only apply to food which is correctly handled before and during the canning process.

So far we have assumed that enzymic damage is of no significance in canning foods because the heating times are far in excess of those required to destroy enzymes. However, during the time taken to pack the food into cans, exhaust the air before sealing, and bring up to temperature in an autoclave, some enzymes, particularly in fruit and vegetables, may be sufficiently active to produce discoloration or 'off' flavours. For this reason most fruit and vegetables are heated at 90 to 100 °C for one to five minutes to destroy enzymes before packing into cans. This process, known as blanching, also helps to remove air from the tissues.

Chilling

Chilling slows up enzymic reactions and inhibits the activity of all but the psychrophilic bacteria. It slows up spoilage but only allows short term storage of most food.

Freezing

Freezing affects enzymes and microbial activity in two ways. The first and most obvious way is the lowering of reaction rates below those of chill conditions.

The less obvious but more important effect is the immobilization of water as ice, so that it can no longer take part in hydrolytic reactions or act as solvent for reactions. Even above the freezing point some water in foods is not available as solvent. This is the water bound to proteins and other macromolecules which, in the case of proteins, is an essential part of their tertiary structure. Water bound in this way is sometimes described as 'unfreezable'. Not all the 'freezable' water in food freezes at the freezing point of water, and below the freezing point water exists in three states, frozen, unfrozen, and bound or unfreezable. Commercial food freezing reduces the temperature of the food to below −20 °C, where only a very small proportion of the 'freezable' water remains unfrozen.

The liquid phase in food is an aqueous solution containing a very great number of solutes. To understand what happens as food freezes we should first recall the freezing curves for pure water and for a simple solution. An example is shown in figure 4.3.

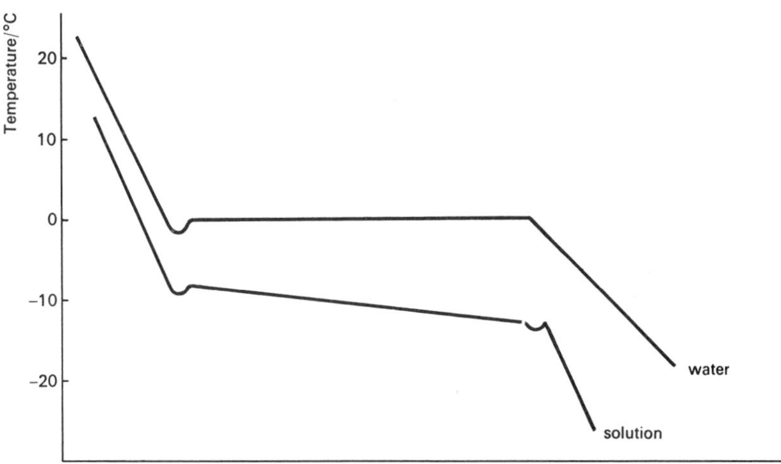

Figure 4.3
Cooling curve for water and simple solution.

As heat is removed from water above 0 °C the temperature drops. If the rate of heat removal is fairly rapid, ice will not normally form at 0 °C and there will be a short period of super-cooling. The extent of super-cooling depends on the rate of cooling and the number of nuclei present to start the process of crystallization. Normally at −1 to −2 °C ice begins to form, the temperature returns to 0 °C, and the ice crystallizes out at 0 °C until the water has frozen. The temperature then drops steadily below 0 °C.

In a simple solution the freezing point will be below 0 °C. Once again there is a short period of super-cooling and a return to freezing point as ice begins to form. As the water crystallizes, the solution becomes more concentrated and its freezing point is lowered still further. The temperature drops steadily until the solution becomes saturated. At this point there may be a short period of super-saturation before any solute crystallizes but after this the temperature rises again to the freezing point of the saturated solution, and the water and solute crystallize out in constant proportions, and at a constant temperature, until all the solution is frozen.

A food can be considered as a mixed solution with solutes A, B, C, etc. in increasing order of solubility. The temperature will behave similarly to a simple solution of A, with ice separating out, until the liquid phase is saturated with respect to A. After this point in a simple solution we would expect the temperature to remain constant while water and A crystallize out in a constant proportion. However, while A is separating out, B and C are becoming more concentrated and the freezing point of their solutions drops. The result of the superimposition of the freezing curves of a large number of solutes gives a curve similar to that in figure 4.4.

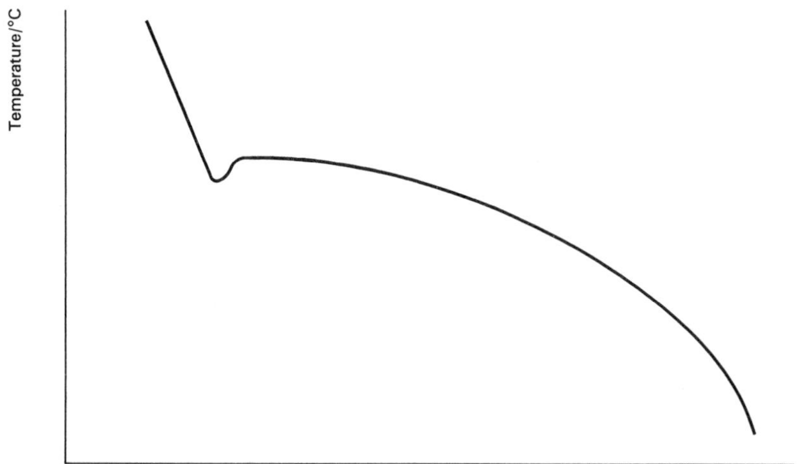

Figure 4.4
Cooling curve for a typical food.

Food starts to freeze at -1 to -3 °C. Each solute separates in turn, when its concentration reaches saturation and when the temperature falls to the freezing point of its saturated solution. The final freezing point may be very low, for example, the freezing point of a saturated solution of calcium chloride is −55 °C and because all foods contain calcium and chloride ions, some part of the water in foods is still unfrozen at −50 °C. Normal storage temperature in commercial practice is −18 to −20 °C and so a small proportion of water (about 2 per cent) remains unfrozen.

The location of ice formation is a critical factor in the quality of frozen food. The aqueous phase in food is within the cells, and in the inter-cellular material. These two locations are separated as long as the cell walls remain intact. The distribution of water between the cells and the inter-cellular material is governed by the osmotic and turgor pressures in those regions. Any changes in relative osmotic pressure will be compensated by movement of water through the cell membrane, but this will take time.

Ice crystals normally start to form first in the inter-cellular material, then in the cells. Consequent ice formation depends on the rate of cooling. If the cooling rate is slow the crystals in the inter-cellular regions continue to grow. The vapour pressure of water is lower at the surface of ice crystals than in the rest of the tissue and so water migrates towards the crystals and diffuses out of the cells into the inter-cellular material. This results in the formation of a relatively few large crystals in the inter-cellular material. If freezing is rapid, the temperature will drop to the freezing point of the liquid phase within the cells before any appreciable migration of water from the cells can occur, and a much greater number of smaller crystals are formed throughout the tissues.

Water can also migrate during thawing, or if the frozen food is stored at too high a temperature. If the food remains for a long time between 0 and −15 °C, the water will migrate and the larger crystals will grow in size.

When food is frozen –

1 The food changes in volume. Water expands on freezing, but the solids contract. This can set up stresses within the tissues.

2 Ice crystals form preferentially in the inter-cellular material. If they grow to any size they cause stresses within the tissues, which may be enough to rupture cell walls.

3 The concentration of solutes increases. The effect is greatest within the cells, particularly if freezing is slow. The changing concentration alters the pH, sometimes by as much as one to two units, alters the ionic concentration of the solution, and alters the osmotic pressure. These changes will denature proteins and change the physical form of the other molecules. This results in a breakdown of the structure of the cells, and irreversible changes in the structure, semipermeability, and water holding capacity of the cell walls.

These physical and chemical changes alter the physical form which gives food, particularly fruit and vegetables, its crispness. Freezing nearly always results in a poorer texture. This is particularly noticeable in some fruits such as strawberries, which after freezing and thawing are very soggy. Lettuce is just not worth eating after freezing and thawing, and in general fruit and vegetables with very high moisture content (85 to 90 per cent) are so damaged by freezing as to be unacceptable. Microscopic examination of slow-frozen plant tissue after thawing shows spaces where large ice crystals have been. When the crystals melt, the water is not reabsorbed into the tissue.

Meat and fish are less damaged structurally but lose a quantity of fluid on thawing. This moisture is unsightly, flavour and nutrients are lost, and there is often an increase in toughness. In animal tissues, which consist of bundles of long pliable fibres, the damage is mainly in the inter-fibre region, resulting in separation of the fibres and damage to the cell contents.

The preceding paragraphs explain why it is important to freeze food as quickly as possible and to store it at constantly low temperatures. The major aim of industrial food freezing is to remove heat from the food as quickly as possible.

The thermal properties of the food set the limits to the rate of freezing. If a medium-sized strawberry is dropped into liquid nitrogen at about −190 °C, which provides about the maximum practicable temperature differential, it still takes about five minutes for the centre of the berry to freeze, and even this rate of freezing causes a lot of damage to the texture of the berry.

Enzymes and micro-organisms also influence the quality of frozen foods. The rate of all enzyme reactions are very greatly reduced in frozen storage and the activity of many enzymes in frozen tissue is undetectable. Some enzymes of vegetable tissue, however, remain sufficiently active to produce 'off' flavours in a few weeks at −18 °C. To avoid this deterioration most vegetables are blanched before freezing, that is, they are heated in steam or in water at about 95 °C for one to five minutes. Blanching causes loss of weight and it is desirable to keep the blanching time to a minimum. In normal commercial blanching the raw material is heated until tests for the enzymes catalase or peroxidase in the product are negative. These enzymes are probably not important in food spoilage but they are more heat stable than most enzymes, are easy to assay, and provide a useful index of adequacy of blanching.

Freezing kills about 50 to 80 per cent of micro-organisms and is most effective when it is slow. Moulds are generally more resistant than bacteria. The majority of surviving bacteria are inactive in frozen foods but there are reports of bacteria growing at temperatures as low as −12 °C and moulds at −9 °C. The surviving bacteria become active and multiply quickly after thawing. In order to avoid microbial spoilage it is essential to store frozen foods at the correct temperature, and they should be eaten as quickly as possible after thawing. Once thawed they should never be refrozen.

Dehydration

Dehydration is the oldest form of food preservation and in many countries food is still dried in the sun. Modern drying techniques fall into two classes, air drying and freeze drying. Air drying is achieved by applying temperatures of 40 to 100 °C in air or under some kind of vacuum. The loss of water causes an increase in concentration of solutes in the tissues and causes damage similar to freezing. In addition it causes heat damage, similar to cooking. The overall effects are irreversible loss of structure and denaturation of protein, and consequently less water is reabsorbed on rehydration than is lost in dehydration. Most tissues shrink and never regain their original volume or tenderness. The heat often causes browning reactions and discolorations. Air-dried food has the advantage that oxygen can penetrate into it only very slowly and it does not usually require expensive nitrogen or vacuum packing. However it is normally hygroscopic to some extent and needs moisture-proof packaging.

The moisture content is normally reduced to 5–6 per cent on drying, because some chemical reactions can take place at higher moisture contents. This is well below the limit required for bacterial growth (about 18–20 per cent moisture) or mould growth (13–16 per cent). The drying process destroys enzymes eventually, but before this happens the food is at temperatures where enzymes are most active. To avoid spoilage during drying most fruit and vegetables are blanched before drying.

Freeze drying is a comparatively recent development. Food is first frozen and then the water is removed by sublimation of the ice under vacuum. This avoids damage due to heat, but freezing damage occurs. The dried product has an open, expanded, porous structure. This rehydrates much more easily than air-dried material and has a better texture and flavour when reconstituted, but its open structure exposes a very large internal surface for oxidation. Freeze-dried material is very fragile and tends to powder during packing and transport.

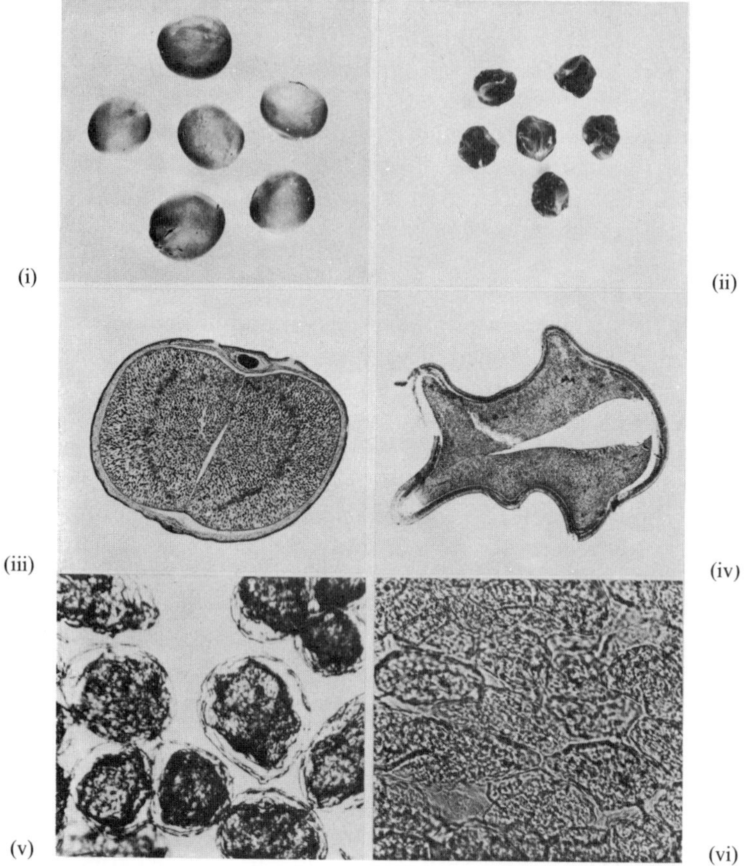

Figure 4.5
Comparison of air-dried and freeze-dried peas.
(*i*) Freeze-dried peas (× 2). These largely retain their original shape, although cracks develop in the skins.
(*ii*) Air-dried peas (× 2). Shows shrivelled appearance due to contraction on drying.
(*iii*) Section of freeze-dried pea (× 10). This shows retention of much of the original cellular structure.
(*iv*) Section of air-dried pea (× 10). Shows contraction and fusing of cells.
(*v*) Section of freeze-dried pea (× 800). Shows cells still capable of separation.
(*vi*) Section of air-dried pea (× 800). Shows dense structure of collapsed and fused cells.
Photos, Unilever Research Laboratory

Curing

Curing is the process of adding common salt to food. It increases the ionic concentration of the liquid phase above that at which micro-organisms can grow and multiply. The practical problem in curing is to achieve a sufficiently high salt concentration to inactivate micro-organisms without making the food

A typical scheme for preparing vegetables for processing is:

Raw vegetable ⟶ screen ⟶ winnower ⟶ washer

processing plant ⟵ sorting ⟵ air blast ⟵

The screen removes leaves and stones. A blast of air winnows the falling mass of vegetables to remove light debris. The second air-blast dries the produce and also serves to cool it, through the evaporation of water from the surface. Sorting may be visual or mechanical. For example, peas are passed along a travelling belt under revolving drums studded with fine bent needles which pick up peas containing worm holes.

There are many ingenious methods for removing peel and the stones of stone fruit. Carborundum-lined drums remove hard skins (as on potatoes) by abrasion; hot water or steam treatment causes soft skins to puff and loosen (as on tomatoes). Stones or cores can be removed by halving the fruit and scraping out with specially shaped knives, or, with small fruit like cherries, by plungers which push the stone out of the whole fruit.

Dairy products

Milk is a good medium for bacterial growth and is readily infected by micro-organisms. Processing destroys most of these micro-organisms but, as we have seen previously, the number which survives is dependent on the original numbers present. For this reason high standards of hygiene are required in the dairy and in transport to the processor. Bovine tuberculosis and *Brucella abortus*, which causes cows to abort and produces undulant fever in humans, are reduced by good farming practice. Increased dairy mechanization, such as the use of milking machines, reduces the danger of contamination by dairymen but may be a source of infection. For this reason equipment must be kept scrupulously clean, and washed regularly in disinfectant.

Frozen liquid egg, spray-dried egg, and freeze-dried egg are convenient substitutes for fresh eggs for the confectioner and baker. Stringent hygiene must be observed both before and during processing. Washing the shells reduces the possibility of infection. The water should be warmer than the eggs, since cool water will cause the contents of the egg to contract and allow the cleaning water to infect them. Contamination of the egg-shell is reduced by keeping the hens in clean accommodation. After breaking each egg, the operator smells it in case it is bad. Even with the best conditions of rearing it is possible that an infection may be passed into the egg from the hen. The *Salmonella* group of micro-organisms can enter the eggs in this way. This infection is common in duck

taste too salty. It is particularly important that the salt concentration should vary as little as possible throughout the food, otherwise one end of a rasher of bacon may taste too salty, while at the other end the bacteria are multiplying rapidly.

Pickling

Pickling is the process of soaking food in an acid solution, normally vinegar, so that the pH is too low for microbial growth or enzyme activity.

Irradiation

Irradiation is sometimes referred to as the preservation method of the future. Food is exposed to γ-radiation from a Cobalt 60 source or to β-rays from a linear accelerator. These destroy micro-organisms and, if the food is exposed for long enough, it is sterile. Unfortunately irradiation often produces a very unpleasant flavour and a cooked texture. The chemical changes occurring are very complex, and very little is known about them at present.

If the radiation is reduced to levels below those which spoil the flavour, the product is not sterile but many of the surface micro-organisms are killed. This process can be used to increase the storage life of fresh strawberries. Irradiation also increases the sensitivity of micro-organisms to heat and so the heat treatment required for sterilization of canned foods can be reduced by previously irradiating the food. The advantages of this process over conventional canning are not great enough at present for it to replace the older method.

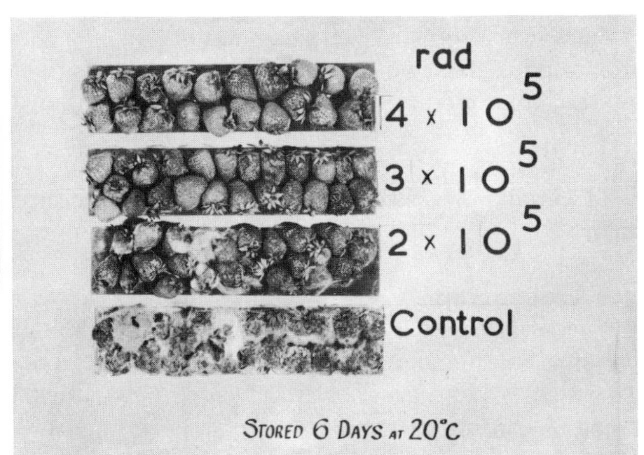

Figure 4.6
Comparison of storage properties of irradiated and non-irradiated strawberries. This illustrates the effect of γ irradiation from a Cobalt 60 source in destroying mould cells on the surface of fruit. *Photo, United Kingdom Atomic Energy Authority.*

Chapter 5

Industrial food processing

Industrial food processing is concerned with applying the basic principles outlined in the previous chapter to the design of the most suitable equipment and processes. Some methods of preservation, for example smoking and sun-drying, have been known for many hundreds of years. Others (for instance freeze drying) have been developed with the advance in our knowledge of food chemistry and physics in the present century. Greater efficiency through automation, increased control, and better quality are sought in existing processes. In the less industrialized parts of the world there must be a different approach. The high cost of many processed foodstuffs, the lack of sophisticated equipment, and the nature of the raw materials and diet need a most imaginative application of our knowledge of food preservation to improve the nutrition of these poorer countries. In these areas, where malnutrition and poverty are common, the fullest possible use of the available resources of food, manpower, and plant is essential.

The industrial food processor is concerned with converting the raw food material into the finished product of the desired quality at the minimum cost. He is dealing with a complex and variable raw material. Modern processing plants are most economical when dealing with uniform material requiring uniform treatment. In the food industry a compromise is needed which involves making the plant as flexible as possible (that is, capable of working with a wide range of raw materials) and, where possible, selecting and grading the food before it is processed. For example, fruit which is ripe and soft may be suitable for making into jam but the firmer produce may be better used for bottling as whole fruit.

With these points in mind we will now consider the major industrial processes used for food preservation.

Preparation for processing

Agriculturalists play an important part in ensuring that the raw foodstuffs will be suitable for processing. Selection and cross-breeding alter the yield and resistance to disease. Resistance to bruising, improvements in flavour, and regularity of size and colour can all be improved. As farms become more mechanized, the produce is subjected to rougher treatment than with slow manual methods. More and more the food industry is setting the targets for the farms; contracts for pea growing for quick freezing specify the variety to be grown, the time of planting, and the fertilizers, and will arrange for tests for maturity.

Fruit and vegetables

Fresh fruit and vegetables in any greengrocer's shop show that they are prone to attack by a variety of insect pests and diseases which may be reduced by the use of pesticides and fungicides before harvesting and by the development of resistant varieties. However, the processor must still inspect and clean the produce carefully; soil, stones, and inedible portions of the food have to be removed.

Mechanical tests have been devised to test the quality of the raw material for processing. In some cases, payment to the grower is based on the measurement of tenderness. The 'tenderometer' is a device used in such a way for determining the best time to harvest peas for processing. A sample of the peas is crushed between two intermeshing metal grids and the resistance of the peas to crushing is indicated on a scale, this being taken as a measure of tenderness. A photograph of a tenderometer is shown in figure 5.1.

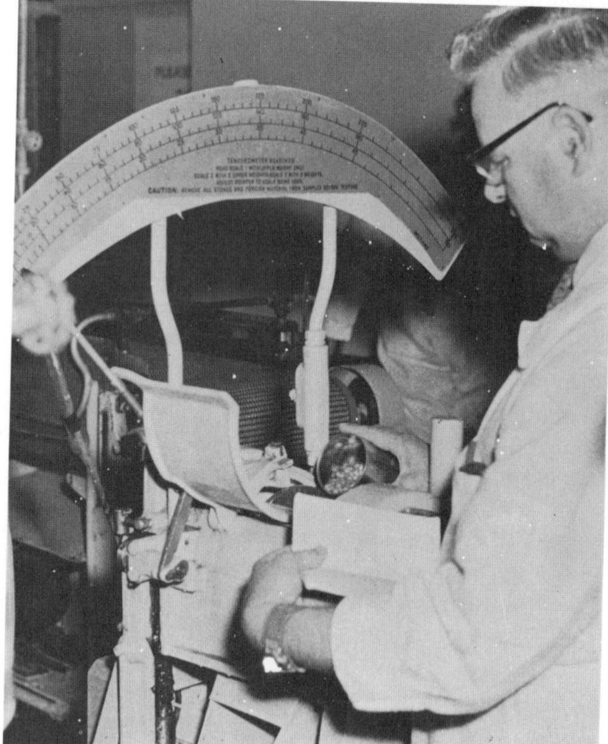

Figure 5.1
A tenderometer for measuring the tenderness of peas.
Photo, Unilever Research Laboratory.

eggs. Some heat processing (or perhaps in future, irradiation) will therefore be needed to prevent contamination of processed egg supplies.

Meat and fish

Meat. The meat processor buys a great deal of his raw material in the form of complete carcasses. Only a proportion of the carcass can be sold as joints of fresh meat, and he must find uses for the rest. Fat, rind, and parts of the meat which are less attractive as joints can be used in sausages. Bones can be converted by high pressure cooking into glue and into bone meal for animal feed or for use as a fertilizer. Some fat may be rendered to produce lard and dripping and some goes to soap manufacture.

Conditions of slaughter affect the quality of the meat produced. Pigs which are excited or frightened before they are slaughtered consume the carbohydrate glycogen in the muscles before they are killed. The lactic acid which would normally be formed from the glycogen after slaughter is therefore not present. This lactic acid appears to play an important part in the quality of the meat, acting as a preservative and improving the quality. Care has to be taken therefore to keep the animals rested before slaughter. The slaughtering area must be kept clean. Possible sources of contamination, such as the internal organs of the animals, must be kept separate from the meat for processing. Trained inspectors check each animal for signs of disease. Diseases such as anthrax have been virtually eliminated in this country by strict hygiene control in slaughterhouses.

Fish. In recent years fishing fleets have had to go further afield for their catches owing to the depletion of coastal waters. This has led to a need for improved methods of keeping the fish whilst it is brought to port. Ships are now equipped to process the fish at sea. Cooling at 0 °C in ice is not sufficient for long term storage. At this temperature the bacterial population increases sufficiently fast to spoil the fish, and causes the formation of trimethylamine which is the main compound responsible for the characteristic smell of 'stinking' fish. With new techniques the cleaned fish is frozen rapidly (to approximately −15 °C) and stored at even lower temperatures (−20 °C or below). In this way really fresh fish can be brought to the housewife.

Cooking and pasteurization

Cooking

Cooking of foods inactivates enzymes and reduces the bacterial load. It is frequently combined with other preserving techniques, such as curing, to improve the keeping quality. For example, cooked hams are produced from cured pork. The meat is pressed into metal moulds (often made of aluminium) and held down firmly by a sprung lid. The moulds are then placed in either hot water or steam. The reduction in weight which occurs during most forms of cooking represents a substantial loss of costly material to the processor and there is evidence that this loss is affected by the time and temperature of cooking. The necessity for rapid cooling has been stressed in the previous chapter. In practice this requires removing the loads of cooked product from the cookers, and immediately placing them under cold water sprays.

Industrial ovens for baking foodstuffs vary in the use they make of the three methods of transferring heat: convection, conduction, and radiation.

In the 'travelling oven' air heated to 200 to 230 °C is circulated around an endless belt which carries the product through the oven. The speed of the belt is such that the baking is just completed at the exit end. Such ovens are frequently split into sections through which the product travels in turn. These sections are heated to different temperatures since the initially moist product may require different baking conditions to those for the nearly baked product. The greater the speed of the forced circulation of the air over the food (by fans) the greater the amount of heating by convection.

Direct-heated ovens are those where heat is transferred directly from heating source to product. For example, the foodstuffs can be placed on a heavy metal plate which slides into the oven-housing where it is heated. In this case much of the heat is transferred to the product by conduction. Radiant heaters are sometimes used to provide a substantial part of the heating requirement by direct radiation, although in all types of oven some radiation is always present from oven walls.

Pasteurization

Pasteurization is the term used for a heat treatment of a foodstuff, which kills or inactivates the heat sensitive disease-producing organisms without achieving complete sterilization. It is applied particularly to milk processing. Improvements in heating methods and equipment have changed the methods of milk pasteurization. The simplest form of milk pasteurization is called the holder-process method. The milk is heated rapidly to 65 °C, held at this temperature for 30 minutes, and rapidly cooled. The heating and cooling is carried out in heat exchangers of the type shown in figure 5.2. Milk passes over one side of each plate and the heating or cooling water over the other, so that a very rapid interchange of heat takes place.

The rapid transfer of heat in such exchangers is the basis of the newer high-temperature, short-time (HTST) pasteurization process. The raw milk passes through the first plates of the heat exchanger and is warmed by the pasteurized milk leaving the unit. The warm milk then passes into the heating section where hot water is used to raise and hold the temperature at 72 °C for 15 seconds. The pasteurized milk is returned to the first section to give up some heat to the incoming milk. The final plates are cold water cooled. For rapid heat exchange the plates require a large surface area, and a high degree of turbulence is necessary in the two liquids. Narrow gaps between the plates increase flow rates and ensure even temperatures throughout the bulk of the fluids.

Ultra high temperatures (UHT) of around 140 °C can be used to pasteurize or even to sterilize milk in a few seconds. With these high speeds the cooked flavour, which sterilization causes when longer processes are used, is avoided. The product can be kept for several weeks and is almost indistinguishable in taste from pasteurized milk.

Figure 5.2
A plate heat exchanger used in the pasteurization of milk.
Photo, Alfa Laval Company Ltd.

Canning and sterilization

The determination of the temperature and time for the canning of a particular commodity has been described previously; factors such as pH, salt content, and initial bacterial contamination affect the heating conditions. Every can must be sterile, so the processor must ensure that there is the minimum variation from can to can. Careful selection and efficient pre-processing of the raw material are therefore essential.

Steam is the most common source of heat for canning processes. At atmospheric pressure steam has a temperature of 100 °C and this is too low to kill the heat-resistant spores of bacteria. Only those foodstuffs such as fruits, pickles, and fruit juices, in which spores cannot grow can be safely processed in cans at atmospheric pressure where it is the acidity that provides the additional protection. We have seen that for those foodstuffs where more excessive heat treatment is required, higher temperatures and shorter times are less detrimental to quality for the same bacterial kill. The temperature of steam at various gauge pressures (pressure above atmospheric) is shown below:

Steam pressure (above 1 atm) /N mm $^{-2}$	Temperature /°C
0.05	111
0.10	120
0.15	128
0.20	134

The 'tin' can is tin-coated steel containing from 0.2 to 2.0 per cent tin. Some foods will bleach in these standard cans, and for these products the tin plate is lacquered on the inside. The cans may also be made from aluminium. Glass bottles have long been popular, particularly where the processed food has an attractive appearance, but some foods may be damaged by light when stored in glass containers. Some light-induced reactions may cause 'off' flavours and loss of nutrients.

In a typical canning process the food is sorted, cleaned, and blanched to destroy enzymes and to remove the air retained in the tissues. The blanching may be by immersion in hot water or by blowing steam through the raw material for up to ten minutes. The clean cans are then filled by an automatic dispenser which puts the correct quantity of material in the can. The remaining air in the can is replaced by steam in an 'exhausting' unit. This forms a vacuum in the can when the steam condenses. The lids are applied by machines which press the lid and body of the can together to form an airtight seal. The cans are then placed in a retort.

The simplest retort is the still autoclave, a pressure-tight heating chamber. Crates of cans are placed in it, steam is blown in, and the lid is closed tightly. After the air has been blown out through a vent the steam pressure is brought up to the required level and the temperature is automatically controlled over the whole process time. Although the still retort is a simple piece of equipment it is limited in that it is unsuitable for processing at temperatures much above 120 °C because the low rate of heat transfer inside the product causes the outer layers to become over-cooked. Retorts have been developed in which the cans are agitated during processing by revolving the cans around the axis of the retort or by reciprocation. The agitation circulates the material inside the can so aiding the action of natural convection.

On leaving the retort the cans are cooled by immediate immersion in cold water. The cooling water must be sterile since it is a source of infection by being drawn into the occasional 'leaky' can as a result of the vacuum caused by cooling.

Refrigeration

Cooling is required after all forms of heat processing, often between the various stages of a process, and for storage of both raw and finished material. Freezing is a special use of refrigeration for long term preservation.

Packing in ice is one means of refrigeration for temperatures above 0 °C. The majority of the heat removing capacity of ice comes from its latent heat. One kilogramme of ice will take up 335 kJ of heat when it melts; when the same weight of liquid water increases in temperature by 1 °C it takes up only 4.19 kJ. Therefore it is necessary to have a constant supply of ice to replace the liquid water formed on thawing. 'Dry-ice' (solid carbon dioxide) is so called because it sublimes directly to a gas on taking up heat. At atmospheric pressure the sublimation temperature is −78 °C. Solid carbon dioxide has twice as much refrigerating capacity as ice for equal weights. For these reasons dry-ice is convenient for use in cooling food transport vehicles, if the length and time of the journey are not too long.

Mechanical refrigeration is required for cold storage rooms, freezing, and most cooling operations as well as for some forms of transport. The refrigerants used are commonly ammonia and fluorinated hydrocarbons. Ammonia is economical and easily detectable by its pungent smell. It can, however, form explosive mixtures with air. Freon 12 is the trade name of the most popular of the fluorinated hydrocarbon refrigerants. Its non-toxicity and non-flammability together with its thermal properties make it an ideal refrigerant.

Liquid refrigerant passes from a receiver, through an expansion valve, and into an evaporator held at a pressure low enough for the liquid to evaporate. The evaporator normally consists of coils or fins exposed to the surroundings to be refrigerated. The latent heat of evaporation required is drawn from the surroundings which are consequently cooled. The refrigerant vapour is then drawn into a compressor which increases the pressure and moves the vapour to a condenser. This is a heat exchanger where the vapour is condensed by transferring heat to a coolant, normally water. The liquid refrigerant passes back to the receiver and the cycle is repeated, at the cost of expending energy in the compressor and withdrawing heat in the heat exchanger.

Whether it is a room or a piece of plant which has to be refrigerated, a vital factor in design is the insulation. Loss of heat to the external atmosphere increases the required capacity of the refrigeration unit. The earliest insulants, such as cork, have to some extent been replaced by fibre-glass, mineral wool, and foamed plastics (polystyrene and polyurethane).

Fruit and vegetables require particular care in refrigeration. They continue their life processes at a reduced rate and the production of carbon dioxide in their respiration evolves heat. Poorly stacked peaches in one cold store have been found to *rise* in temperature from 10 to 30 °C in 150 hours due to the heat of respiration and microbial activity. The replacement of part of the oxygen in the atmosphere of the cold store by carbon dioxide is sometimes used to slow down the metabolic processes of fruit and vegetables. Temperatures during long-term storage require careful control as all fruits are damaged by freezing and some (for instance bananas) by holding at temperatures above this.

Freezing

We have seen that rapid freezing is essential to prevent tissue damage and ensure the quality of frozen foods. There are three general methods whereby this can be achieved. These are:

1. Freezing in air.
2. Freezing by contact with refrigerated surfaces (for instance, metal plates).
3. Freezing directly in liquid refrigerants.

In all these methods of freezing the initial preparation of the food is important. Enzymes which would survive freezing and affect storage have to be destroyed, usually by blanching. It may be necessary to cut the food into a shape suitable for freezing. Variations in the thickness of the portions can lead to under-freezing, or can slow down the freezing rate. In many cases the product is packed before freezing, and portions of the correct size are required for filling the cartons properly.

Packaging or covering stops the foodstuff drying out during freezing. Water vapour in the air of the freezing compartment freezes out onto the surfaces which are cooling the chamber. This reduces the water vapour in the air and leads to dehydration through evaporation or subliming of water from the product. When this happens, whitish dry areas, due to a condition known as 'freezerburn', may develop in the foodstuff. There is a serious deterioration in appearance and quality when this occurs and the changes which take place are mainly irreversible. For example, fish develop a matt appearance and woody texture.

The production of fish fingers provides a good example of the preparation of food for freezing. Fish fillets are packed into shallow cardboard boxes (approximately $60 \times 30 \times 4$ cm). These are frozen in a plate freezer and cut into 'fingers' by a mechanical saw. The pieces pass through a tunnel where they are sprayed first with a batter of egg, flour, and milk, and then with breadcrumbs. The coated fingers are fried in fat in a stainless steel bath and then packed ready for refreezing in a plate or air freezer.

If the food is cooked before it is frozen it will be ready for eating when thawed. Complete meals can be frozen in this way and stored under refrigeration until required. This is a convenient method for preparing meals for large scale catering in schools or hospitals.

Freezing in air

In still air it is only necessary to provide a suitably insulated chamber and to hold the stacks of foodstuff in it until they are frozen, but this takes a long time and as we have seen, a slow rate of freezing usually spoils the food.

Increasing the air velocity over the product increases the freezing rate. At -30 °C and a velocity of 15 m s^{-1} the fish fillet blocks mentioned above would freeze in about four hours. With a velocity of only 2.5 m s^{-1} this time would be approximately doubled. Hence for the quickest freezing refrigerated air is blown over the food at high speed through narrow chambers or tunnels. Such units are called blast freezers and they are used for freezing anything from cans of liquid egg to joints of meat.

An increase in air velocity also increases the drying rate. Covering the food and keeping a high air humidity greatly reduces the amount of moisture lost in this way.

Plate freezing

The plate freezer consists of a refrigerated chamber containing refrigerated metal plates between which the product is frozen. The refrigerant passes to each plate, generally circulating by means of coiled pipes. Or the refrigerated surface may be a continuously moving belt on which the foodstuff is carried in and out of the freezing compartment. A high freezing rate depends on good contact between foodstuff and plate. Therefore the food should have large flat surfaces; peas and fish fingers, for example, are packed in rectangular cartons.

Immersion freezing

The foodstuff is plunged directly into a bath of liquid refrigerant. Salt brines were originally used for fish freezing as they are convenient and cheap. They cannot, however, be used with fruit and vegetables, but invert sugar and glycerol solutions have been used for these. Packaged fruit juices may also be frozen by this method.

A spray of refrigerant is an alternative to the liquid bath. The latest development is to use liquid nitrogen for this purpose. The very low temperature gives high freezing rates and makes available better quality frozen soft fruits (raspberries, strawberries).

Dehydration

Dehydration processes are based on the evaporation of water from the foodstuff and the continuous removal of water vapour from the environment. Air is the usual drying medium as it is plentiful and easily controlled. The air can be used both to supply the latent heat of evaporation and to carry away the evaporated moisture. Alternatively, the food can be dried by placing it in contact with a heated surface and removing the vapour by applying a vacuum. In the initial stages of drying, the water is easily removed from the food since it travels from the interior to the surface fairly readily and then evaporates as if from a free water surface. The movement becomes progressively more difficult as the moisture content falls. The change in drying rate is illustrated in the dehydration of potato slices. The water content can be reduced from 4 to 0.5 grammes of water per gramme of dry matter in about one hour, but a further reduction to 0.05 grammes can take eight hours.

The initial preparation of the food for drying is similar to that for freezing, but the raw material usually has to be cut or sliced to make drying easier. Potatoes are cut into strips or diced; meat and fish are cut into small pieces, cooked, and minced down to approximately 0.5 to 1 cm. Machines have been developed which prick peas to allow easier transfer of water from the centre, and cut sprouts before drying to allow the moisture to leave the centre without having to pass through the tissues.

Pre-cooked dried foods (such as chicken and vegetable pieces) are commonly mixed together to form complete dried meals. They have a long storage life and require little preparation by the housewife.

With liquids such as milk it is normally more economical to remove some of the water by evaporation before drying. This is done by boiling the milk under vacuum. A pressure of 1.6×10^{-2} N mm^{-2} reduces the boiling point of milk to about 55 °C and at this temperature there is little development of a 'boiled' flavour.

Air drying

Heated air is blown over the foodstuff at rates of up to 5 m s^{-1} either in a tunnel, through which the produce is conveyed on shelved trucks, or in a cabinet fitted with perforated trays. Diced vegetables and fruits are dried in this way.
If drying air is blown upwards through a bed of food particles (peas, potato granules) the particles can be separated from each other and a large surface area is exposed for drying. At a certain air speed the bed becomes 'fluidized' and will flow in a similar way to a liquid. High drying rates and continuous operation are made possible in this manner. Continuous operation is also achieved by blowing the air through food pieces carried on open mesh conveyors.

Spray drying

The spray drier is an air drying plant for liquids. Figure 5.3 shows such a drier. The principle of operation is simple. The liquid is sprayed vertically or horizontally into heated air in a chamber; this evaporates the water to give minute dry spheres of the solid material. The drying air, heated outside the chamber, flows either in the same or the opposite direction to the spray. The majority of the drying takes place before the particles reach the wall, where they are liable to stick if they arrive while still wet, so the chambers must be large, often up to 15 metres high. The dry particles fall to the base of the chamber where they are drawn out and collected.

The process gives extremely rapid heat exchange and little change in flavour. The product is highly soluble as can be seen from the ease of reconstitution of 'instant' coffee. Spray-dried milk is prepared from liquid evaporated to about 60 per cent moisture content. A secondary process is used to give more ready reconstitution with water. This requires re-wetting of the dry particles, e.g. in a vertical tower down which the dry powder falls through a water saturated atmosphere. The particles cluster together, are removed from the tower, and re-dried to about 4 per cent moisture. There is much less tendency to form hard lumps when water is added.

Figure 5.3
A spray drier.
Photo, Niro Atomizer A.S.

Drum drying

Liquids and pastes can be continuously dried into flakes by the application of heat through a metal surface. The drum drier consists of a metal drum heated internally by high pressure steam. The drum is rotated slowly whilst the fluid food material is deposited on it in a thin film. The dried product is scraped off the drum by accurately positioned blades. In drying milk (from about 82 per cent moisture content) the liquid is fed from a trough onto the drum. The product has more of a 'burnt' flavour than that produced by spray drying. Mashed potatoes are dried to potato flakes by spreading a slurry of the cooked potatoes on the drum.

Freeze drying

This process is at present the most expensive in commercial use for drying but it gives, for some foods, the highest quality product. The material is frozen and placed on trays in a cabinet attached to a powerful vacuum line. The process is based on the sublimation of ice to vapour, giving a minimum amount of shrinkage and a highly porous product. For water to exist only in the solid and vapour states at 0 °C the pressure must be below 6.1×10^{-4} N mm^{-2}. As we have seen, low freezing point mixtures are formed in foodstuffs and therefore in practice freeze-drying requires pressures of 1×10^{-5} to 3×10^{-4} N mm^{-2}.

In the standard freeze drying process the product is held between flat metal plates during drying. This enables the latent heat of sublimation to be supplied, but the drying rate is retarded by the lack of channels for the escape of water vapour. Consequently, accelerated freeze drying (AFD) has been developed which uses expanded metal mesh on either side of the product to give vapour channels. Improvements in transferring heat from plate to foodstuff have been achieved by using plates with spikes which protrude into the food. The porous freeze-dried food may, especially if fats are present, deteriorate in the presence of oxygen. So the vacuum in the freeze-drier chamber is broken with nitrogen and the product is packed in nitrogen.

Curing and pickling

In curing and pickling, salt or acid reduce unwanted microbial growth. They also control the growth of other, harmless, micro-organisms which produce characteristic flavours.

A typical meat curing solution contains 25 per cent salt, 1000 parts per million sodium nitrite, and 1 per cent potassium nitrate. Sugar may also be added. Some of the bacteria from the meat are capable of reducing the nitrate to nitrite. The myoglobin in the meat is converted by nitrite ions to nitrosomyoglobin, which gives the familiar red colour to the product. Quicker curing is achieved by injecting the brine through hollow needles into the joints of meat. This is the first step in bacon curing, followed by holding in a curing solution for about four days.

Some cured products (such as herrings which have been soaked in a strong salt solution) are exposed to smoke from a wood fire. Smoking gives a good flavour and colour to the food, as well as producing a surface coating which reduces microbial attack.

Vegetables such as gherkins and onions can be preserved by pickling. The vegetables are first soaked in a salt solution for several days, while acid is produced by the action of bacteria on the carbohydrates. The salt is then leached out by washing in warm water. The resulting material is left in a weak vinegar solution until it is packed, again in vinegar.

Irradiation

High cost of the plant and the 'off' flavours produced limit irradiation as a means of food processing, though it may be used in combination with other forms of processing (for instance, heat treatment or freezing), or where only fairly small radiation doses will achieve the required purpose.

Ionizing radiations for food processing are obtained from a machine which produces a high voltage (up to ten million volts) to give a stream of electrons which bombard the food. Or a radioactive material can be used, for instance cobalt 60, which emits gamma radiation. Although gamma rays penetrate further than electrons, they are less controllable. The intensity of the radiation is also very much lower.

The radiation dose used depends upon the degree of processing. The unit of dose is the rad, corresponding to the absorption of 10^{-5} joule per gramme of irradiated material. A dose of five million rads (Mrad) will sterilize to about the level achieved in normal canning. At this level unpleasant flavours develop in the food. Doses of 0.1 to 1.0 Mrad give a 'pasteurizing' effect. Thus a dose of 0.3 Mrad significantly reduces the bacterial load and so increases the quality of fresh fish. 0.5 Mrad destroys the *Salmonella* organism likely to be found in liquid egg. Mould growth on fresh fruit can be suppressed by doses of about 0.2 Mrad.

Very useful results can be achieved by even lower radiation doses. A dose of 20000 rad will eliminate insects from stored grain. The difficulty here is the cost of exposing large quantities of material in portions of a suitable size for treatment. A dose of 8000 rad is used commercially to prevent sprouting in vegetables (particularly potatoes). It is worth noting that a dose of 700 rad is lethal to man.

It is important to remember that, as with opened canned foods, recontamination will take place unless the food is covered after treatment. The standard tin-coated cans are suitable containers, although some interior enamels may be unsatisfactory. Packages made from special laminated foils and plastics resist damage both from handling and from normal radiation levels.

Preservatives

The addition of chemicals to food is strictly controlled by law. Only certain preservatives are allowed, up to prescribed levels. When correctly used, however, such chemicals do have a useful preservative effect.

Sulphur dioxide, generally derived from sodium sulphite, is commonly used. It is added to the water used in blanching fruit and vegetables before drying. It makes the colour better and retains more of the vitamin C. The sulphur dioxide is nearly all eliminated when the food is reconstituted. It is also used to preserve fruit to allow jam-making out of season. During the jam-making it is almost all boiled off.

Benzoic acid is effective against yeasts and moulds and is added to fruit juices and acid foods such as pickled products. Sorbic acid is permitted in cheese and flour confectionery. In the latter case, addition of the acid to the ingredients used for cakes prevents mould formation. Similarly, propionic acid prevents the formation of 'rope' in bread. This is a wet stringy condition in the bread caused by the growth of the micro-organism *Bacillus mesentericus*.

Antibiotics are not usually added to food. One danger is that the bacteria which they are meant to destroy may develop a permanent resistance. This is of importance if the bacteria are disease-producing since it would make medical use of these antibiotics ineffective. However, a naturally occurring antibiotic, nisin, which has no medical uses, is at present used in some cheese and canned foods.

Preservatives which have been traditionally used (e.g. the nitrates and nitrites in curing, and acetic acid in pickling) are also allowed, but are carefully examined from time to time to make sure that they are still safe to use.

Antioxidants

This group of compounds reduces the oxidation of the fatty acids in processed foods. The unsaturated fatty acids, such as linoleic and linolenic, are particularly readily oxidized. Antioxidants are added to food such as oils, fats, and butter. The antioxidants permitted in this country are propyl, octyl and dodecyl gallate, butylated hydroxyanisole (BHA), and butylated hydroxytoluene (BHT). Their action has been found to be improved by compounds known as 'synergists'. Citric and ascorbic acid are typical synergists. Their action seems to be in forming complexes with metals (chelating) which would otherwise promote oxidation.

Hygiene

It should be clear from the previous sections that hygienic conditions of manufacture are an essential part of good food processing. However, the subject is so important that some comments on how this can be achieved are worth while.

Factory

The design of the factory and the way it is maintained can both affect hygiene. A clean bright building is much more conducive to hygienic working than a dark and grimy one. In addition some features of design have a direct bearing on hygiene; the cooling water used to reduce the multiplication of bacteria in a cooked product will itself become a source of contamination unless good floor drainage removes this danger. Inadequate cold storage space may mean that some products are left out in a warm atmosphere for long periods, giving ideal conditions for bacterial growth.

Personnel

Human beings always carry bacteria which may be transferred to the food. Regular washing of hands and arms, clean working clothes, and a minimum of direct handling of food reduce the chances of infection. No one must be allowed to work when suffering from boils or open cuts on the hands. There are legal regulations covering many of these matters.

Equipment

There are two categories of equipment. There are the small items, such as knives and containers, which are easily cleaned in a solution of sodium hypochlorite. In many cases automatic washing machines are used, which spray jets of water over the equipment and then dry it in hot air.

Larger equipment is generally immobile. Ease of cleaning in this case depends on design. Most machines can be stripped down to their component parts, each one of which can be cleaned in a sterilizing solution. They are designed so that both the stripping down and refitting can be done speedily and are usually made of stainless steel or aluminium.

By one technique the plant is not stripped down, but 'in-place' cleaning allows a sterilizing solution to clean all parts. For example, an enclosed tank would be fitted with permanent pipework through which the solution is pumped and sprayed into the interior. This method of cleaning is particularly suited to equipment which deals with liquid foodstuffs such as milk.

Figure 5.4
A milk bottling plant. Note the use of automation, and the cleanliness of the plant.
Photo, United Dairies, Ltd.

The future of food processing

Food must be related to the needs of the people who consume it. Since the majority of the world population does not at present get adequate nourishment there is an urgent need for greater exploitation of our present food sources and the development of entirely new ones. The means by which this may be achieved in the future are considered in chapter 6.

The affluent nations will continue to demand improvements in the quality and convenience of processed foods. The processor will exert greater control over agricultural methods in order to obtain produce which is more consistent in shape, texture, and so on. Food scientists will learn more about the nature of food and the changes produced by processing, and will improve the flavour and aroma of processed foods. Ultra high temperatures in sterilization and very low temperatures in freezing will be more commonly used in the future. Structural damage to the food will be reduced by improved techniques and should increase the popularity of dried foods.

The food industry, dealing with complex raw materials, has tended to cling to traditional methods but more and more foods are now being prepared by automation. We can expect this trend to continue in the future, with the food technologist playing a leading role in the design and control of the manufacturing process.

Chapter 6

Food legislation and public health

Primary factors in the health of any community are the quantity and quality of the food which is available to it together with the way in which food consumption varies among the sections of the community. The special needs of children, expectant mothers, and old people may well not be met in conditions where the normal adult diet is quite sufficient to maintain good health. It is no longer acceptable in a civilized community that some shall starve in the midst of plenty and there is a growing awareness in the world community that the well-fed western nations cannot stand by while millions are undernourished in the developing countries.

These general points need to be kept in mind when we consider, in this chapter, the ways in which legislation regulating the methods of production and the sale of food operates. A vocal minority, especially in Europe and America, opposes the use of additives in foods which preserve them or ease manufacture and the same group often also oppose the employment of artificial fertilizers on the land. The claim is made that these aids in food production and processing are a risk to health. Despite every precaution in making tests, a mistaken judgment might allow the use for a period of a substance which has some rather obscure adverse effect on consumers. But the alternative of allowing no aids at all would ensure starvation for many who can now be fed and would also give a poorer quality of product over the whole range of raw and processed foods.

A further fallacy which also receives wide publicity is the view that all substances naturally occurring in plants and animals are harmless while all 'chemicals' (which would presumably include common salt) are harmful. By experience, which must have included a number of cases of illness or death, certain berries, fungi, fish, and so on have been recognized as harmful. It is clearly unlikely that the plants and animals which have been selected to be grown over the centuries for food will include those which have marked toxic effects. But some will contain chemical substances, naturally occurring in them, which would be harmful if taken in large quantities. Even so essential a food constituent as vitamin D can, if taken in excess, cause illness and even death, especially in children. The toxic properties of the liver of the polar bear arise from its high vitamin D content which the bear is able to utilize as a reserve store of the vitamin.

The purpose of food legislation is to protect the health and pocket of the consumer, by preventing the sale of adulterated, impure, or low quality food, and by controlling labelling and advertising. It is normally possible for the buyer to judge the quality of raw foods but with canned and packaged foods he must rely on the label and to a lesser degree on the information given in advertising. Regulations specify the minimum values of important constituents (e.g. meat content in meat products, fruit content in jam, fat content in cream) and list substances which must not be present or added.

The view is often expressed, especially by old people whose sense of taste has deteriorated, that the food today is not what it used to be. There was, so they say, a golden age when food was free from additives and was produced in natural surroundings. But a brief account will show the transformation which has occurred since the first Food Act of 1860. Gross adulteration and the use of poisonous additives have vanished; the industry is now free to use harmless and necessary aids.

Food adulteration

Up to the eighteenth century, most people in England lived either on the food they grew themselves or on that which their immediate neighbours grew; adulteration was only of importance in costly items such as imported spices, in ales and wines, and in tea and coffee. London, as the one large city, offered the greatest scope for attempts to swindle the purchaser but the city companies, fishmongers, salters, grocers, and bakers, appointed inspectors to exert a restraining influence. Control was restricted to detection of the more obvious additions and to the condemnation of products of low standards; there were no proper test methods, owing to the primitive state of chemistry.

In the early years of the nineteenth century London and the industrial towns of the North and Midlands grew rapidly. Millworkers, potters, metal workers, and other town dwellers depended on cheap food. The absence of a proper form of control of food, especially in the new industrial towns, provided every opportunity for fraud, from deceptions such as addition of harmless arrowroot or flour to thicken cream, to the use of poisonous copper and lead salts for colouring. At the same time that adulteration became widespread, the development of analytical chemistry cleared the way for future control.

In 1850 an Analytical and Sanitary Commission was set up by the famous medical journal, the *Lancet*, to examine the quality of food and drink. Analyses were carried out by Dr Hassall by microscopic examination and such simple chemical tests as were available. The results of this work were a main factor leading to the Act of 1860, which provided for the appointment of analysts to detect adulteration and for the prosecution of offenders for offences under the Act.

This first Act was not effective firstly because authorities, under trade pressure, would not enforce it and secondly because there were few means of detecting additives accurately. Today legislation is more effective and an increasing number of good analytical methods are available – paper chromatography to separate artificial colours, gas–liquid chromatography to determine the residues of pesticides, spectrometry and polarography to determine trace metals, and many others. Yet no analyst who is presented today with a can of meat soup can say with certainty how much meat it contains, for his estimate of meat protein is confused by the presence of milk and cereal proteins and even by added flavouring agents (e.g. sodium glutamate). Still less can he be certain what particular meats have been used.

The Society of Public Analysts was formed in 1875 to carry out the analyses and other examinations required to enforce the law. The public analysts were among the first professional scientists and many distinguished chemists received their first training as apprentices in their laboratories. Not all food companies, even in the 1860s, wanted to deceive the public, and one result of the public analyst's activities was the employment of chemists by food firms to ensure that their products reached the standards required. These scientists were often trained in the public analysts' laboratories. In industry, they examined raw materials and even, though here they often met opposition, studied the processes being used. For instance, as early as 1882 Dr Paul Vieth reported on the researches he was carrying out for the Aylesbury Dairy Company. There was parallel work in many other branches of the food industry, although often with less freedom to publish the results. So began the long and still incomplete transformation of the food industry from craft to modern scientifically controlled technology.

The Food and Drugs Act 1955

The opening paragraphs of this Act prohibit the use of ingredients, additives, or processes which 'render the food injurious to health, with intent that the food shall be sold for human consumption in that state'. It also prohibits offering for sale or advertising such foods for human consumption. It states 'If a person sells to the prejudice of the purchaser any food or drug which is not of the nature, or not of the substance or not of the quality, of the food or drug demanded by the purchaser, he shall . . . be guilty of an offence.'

These provisions give basic protection to the consumer. Some regulations made under the powers of the Act define requirements for the composition of particular foods, including what may or may not be added. Other regulations specify the form of label, the limits of advertising, and the hygiene required at all stages of transport, processing, storage, and selling of food.

Food laws, like most other laws in this country, have grown piecemeal, so that we find provisions for some foods actually included in the 1955 Act and given in detail. Thus for historical reasons, milk, dairies, and cream, occupy a prominent place in the Act. So we find in the Act the explicit prohibition 'No person shall add any water or colouring matter, or any dried or condensed milk or liquid reconstituted therefrom, to milk intended for sale for human consumption.' At the time the Act was passed there seemed no possibility that so reasonable a provision could ever be called in question. Today this clause prevents the use in the United Kingdom of one new process for preparing ultra heat treated (UHT) milk (which will keep several weeks). In this process the milk is rapidly heated by the latent heat of condensing steam and as rapidly cooled by the evaporation of the same quantity of water. Although the final water content of the milk is unchanged, the addition of water at the earlier stage by condensation is an illegal process. Only an amendment of the Act will permit the use of this technically sound process as an alternative to indirect heating which is at present used for UHT milk in Britain. The use of regulations to give similar controls for other foods makes change easier, although even then the time needed is often considerable.

Regulations should protect the consumer, while not needlessly making difficulties for the manufacturer. The Ministers of Health and Social Security, and of Agriculture, Fisheries, and Food are advised by the Food Standards Committee and the Food Additives and Contaminants Committee, which weigh the evidence provided by industry, public analysts, consumer organizations, and expert advisers. The officials at the Ministries consider further views which may criticize the conclusions of their reports and at the end of a long consultation process, regulations and orders are made under the powers of the Food and Drugs Act. A few typical examples will be considered.

The regulations are enforced by the larger local authorities – county boroughs, some urban district councils, and the county councils. Officers obtain samples for test and submit these to a public analyst to examine and analyse. Often, if an isolated sample does not conform to the appropriate standard, discussion with the manufacturing firm is sufficient to cause a change in practice or an improvement in control. There are always powers to bring a case to the courts for an offence under The Food and Drugs Act.

Regulations concerned with composition

The Food Standards (Ice-Cream) Regulations 1959 may be regarded as typical of regulations primarily concerned with composition. They apply to ice cream whether sold as such or as part of a composite food. The standard of composition provides for a minimum of 5 per cent fat and 7.5 per cent of milk solids (protein, lactose, etc.) other than fat. For the product described simply as ice cream, vegetable fat may be used, but for Dairy Ice Cream, Dairy Cream Ice, or Cream Ice, it must be a minimum of 5 per cent milk fat, the milk solids other than fat being 7.5 per cent as before. No artificial sweetener may be used.

Regulations for composition cover a wide range of foods from bread and flour to curry powder. The flour regulations are of particular interest since they impose on the milling industry the requirement to ensure a sufficient content of iron, chalk (as a source of calcium), vitamin B_1 (thiamine) and nicotinic acid (another of the B group vitamins) in the flour. The National flour made during and after the Second World War utilized 80 to 90 per cent of the wheat from which it was milled and contained enough of the B group vitamins and iron. White flour, using only 70 per cent of the grain, is required to make white bread of a colour acceptable to the British public, but this contains less vitamins and iron; so these two vitamins and iron are added. The regulations also control the bleaching and improving agents which may be added to flour, and they prohibit the addition of any colouring matter except caramel.

Regulations concerned with additives and contaminants

There are two ways of attempting to control substances which may be added to food to serve some useful purpose but which are not themselves foods. The first is to give general permission but to prohibit certain substances thought to be harmful. The second is to have an exclusive list of additives. Tests on animals (rats, mice, dogs, and even pigs and primates) are used to try to detect whether a substance affects growth, reproduction, the structure of various organs, the prevalence of cancer, or to find whether the substance appears to accumulate in the animal; this guides experts to judge the risk in using a particular additive. In the United Kingdom at present there are lists of permitted preservatives, solvents, colours, emulsifiers, and antioxidants, but it has not yet proved possible to control flavours in this way.

Other regulations are concerned with substances which no manufacturer would wish to add to foods deliberately but which can arise from the chemicals used in agriculture, from machinery, or from other sources. The Arsenic in Food Regulations set limits of 0.1 parts per million for the arsenic content of drinks and 1.0 parts per million for almost all other foods. Fish (including crustacea, molluscs, etc.) and seaweed are exempt from the provisions of the regulation because 'arsenic in proportions exceeding one p.p.m. is naturally present'. The Mineral Hydrocarbons in Food Regulations are designed to prohibit the use of paraffin, etc., as deliberate food constituents; trace contamination from lubricated machinery is not harmful and so is not prohibited – it would indeed be difficult to prevent completely.

The use of gamma rays and high speed electrons to sterilize or 'pasteurize' foods results in chemical changes which must be examined in much the same way as food additives. A new committee will advise on irradiated foods. If satisfactory evidence can be given of the absence of harmful effects this committee can recommend that the general prohibition on irradiated foods may be lifted for particular foods and irradiation dose levels.

Regulations concerned with hygiene

The Food Hygiene (General) Regulations 1960 cover a wide variety of provisions to ensure the cleanliness of equipment, premises, and of persons handling food. The provision of conveniences, washing facilities, first aid, and lockers for clothing is required. Further clauses seek to prevent foods in which bacteria grow rapidly from being kept between 10 and 63 °C, the range in which food poisoning organisms are liable to multiply. There are also special regulations for stalls and vehicles used to sell or deliver food.

Food poisoning

The Food and Drugs Act requires any doctor who suspects a case of food poisoning to notify the medical officer of health of his area. Nowadays food poisoning very rarely results from a poisonous chemical in the food but in almost every case the poisoning is due to bacteria or the toxins which bacteria have produced. The medical officer of health then has powers to investigate and to prevent the sale of food suspected of causing food poisoning.

The growth of canteens, school meals, and the habit of eating in restaurants, has multiplied the dangers of food poisoning. Roughly three quarters of the outbreaks occur from meat products, which provide a very good environment for bacterial growth, and in cream cakes, custards, trifles, fish, and duck eggs. *Salmonella* is the most prevalent genus of bacteria and the symptoms it produces are sickness and diarrhoea. There are some 700 varieties of *Salmonella* which can be identified. This is often a great help in locating the source of a food poisoning outbreak. Other organisms causing food poisoning are *Staphylococcus aureus, Clostridium welchii*, and *Clostridium botulinum*. The toxin produced by *Clostridium botulinum* is often fatal if present in an appreciable amount, but the efficiency of commercial canning reduces the risk of contamination. It is largely confined to home preserved meat and vegetables, and the incidence of botulism is now very low in the United Kingdom.

Really strict application of hygiene regulations would eliminate almost all risk of food poisoning, except perhaps from *Salmonella*. Even for this group the incidence would be greatly reduced but the many sources of salmonellal infection – raw pet foods, animal feeding stuffs, human carriers, processed eggs, and meat – make total elimination difficult.

International standards

In 1962 the two United Nations agencies, the World Health Organization and the Food and Agriculture Organization, set up the Codex Alimentarius Commission, with a procedure for the preparation of standards for commodities, labelling, hygiene, additives, pesticide residues, and analysis. The necessarily complex procedures will take many years and even then will have to be fitted into the various forms of national food legislation. But the Codex can lead the way to one set of world standards as the ultimate aim and final achievement.

Some suggestions for further reading on this subject are given at the end of this book.

Chapter 7

The world food problem

The population of the world in 1967 was about 3700×10^6 of whom two-thirds did not get adequate nourishment.

The majority of undernourished people of the world are in the Near and Far East, Africa, and South America; their greatest deficiency is of proteins which contain the essential amino acids. In 1961, one-third of the world's population got an average of 44 grammes of high quality protein per person per day, but the remaining two-thirds got an average of 9 grammes per person per day.

The world population is increasing at the rate of 70×10^6 people per year, and the rate of increase is itself increasing. In the year 2050 the world's population may be about $10\,000 \times 10^6$, and in order to support these numbers even at mere subsistence level it would be necessary that the intensity of Japanese agriculture should be applied to all cultivable land in Asia, and that of Holland applied to all other cultivable land in the world.

Therefore there is an immediate need to use the scientific and technical knowledge now available to produce more food and especially more high quality protein. This could be done in three principal ways.

1 By increasing conventional food sources, both by traditional and unconventional means.

2 By developing entirely new food sources.

3 By eliminating the wastage of food.

Although it has been generally accepted that only protein from animal tissue contains the amino acids essential for human nutrition, in fact the proteins of all actively metabolizing tissues whether of animal, plant, or microbial origin, have the same overall content of essential amino acids. Thus, for example, the proteins of seed germs, of young grass, of micro-organisms, and of meats, all provide the same nourishment.

Increasing conventional food production

Among the many factors which affect the growth of conventional foods, water is of prime importance. Only 11 per cent of the world's cultivated areas are irrigated. This figure could be relatively cheaply increased to 14 per cent and more expensively to 20 per cent.

In the long run desalination of sea water is most likely to meet the needs for increased food production. The necessary energy could be obtained from nuclear power, and in areas of high sunlight intensity it could be obtained from direct solar energy. Widespread irrigation of uncultivated areas such as deserts is unlikely, however, because most deserts are above sea level, and the energy needed for pumping water, even if the water were available, would lead to a world fuel shortage.

Mineral fertilizers, in particular the elements nitrogen, phosphorus, and potassium, also limit the production of conventional foods. Mankind has been steadily using non-renewable ores for thousands of years. It is true that there is an 'inexhaustable' supply of nitrogen in the atmosphere, but it would involve an enormous amount of energy to produce from it the world requirement for nitrogen which is estimated for AD 2000 as about 60×10^6 tonnes per year.

There is no doubt that very substantial increases in conventional foods are possible by applying scientific knowledge of the many factors which affect plant and animal growth. Plant breeding has succeeded in producing varieties of wheat capable of growing in colder climates, of resisting disease, of giving much higher yields, and of producing protein of higher nutritive value for the human consumer. Similar benefits in yield, hardiness, and nutritional content have been achieved with fruit and vegetables.

The rearing of animals may be a wasteful way of using cultivable land, but there may be some further short term exploitation of this source. One possibility is the deliberate encouragement of large herds of game animals in Africa such as zebra, wildebeest, impala, and eland. Eland would be especially valuable since it is easily handled and puts on fat to give good quality meat. This would also apply to antelope in the USSR.

There could be a considerable increase in animal protein by development of modern methods of catching fish. The quantity of fish which the sea can sustain is determined by the availability of plankton, which in turn is controlled by sunlight and nutrients, but there could be a two or threefold increase in present landings. The fish harvested can be increased by exploitation of new grounds (for instance off the western coast of South America), by better means of shoal detection, such as echo-ranging and radar, and by the use of larger trawlers capable of fishing in all weathers, equipped with automatically controlled nets and factory areas for gutting and/or freezing fish as these are caught. This allows a long period of preservation between arrival at the fishing ground and return to a distant base.

While man has no positive control over the growth of fish in the sea, fish culture in ponds has been practised for over 4000 years. More recently, in Japan and

Indonesia, intensive fish and pig rearing enterprises have yielded 300 kg fish and 3000 kg pig meat per 4000 m^2. The efficacy of fish culture is high; whereas 1 kg of fish body protein requires about 20 kg of feed, 1 kg of beef body protein requires 50 kg of feed. In Britain, warm effluent waters from power stations have been used to promote the growth of fish in restricted areas. The White Fish Authority breeds sole, plaice, and oysters on the Isle of Man. When about 2 cm long they are transported to Hunterston nuclear power station and reared in ponds of warm water, where they reach marketable weight in two years instead of four. Salmon, trout, mullet, and lobster are also being bred and reared under artificial conditions of controlled salinity and temperature in Scotland, Israel, and the Channel Islands. The use of heat prevents the winter growth lag.

As a source of protein, plants are more productive than animals. The area of land required to produce 20 kg of protein per year is 400 m^2 for legumes, 1000 m^2 for cereals, and 4000 to 10 000 m^2 for meat animals. Moreover, plant proteins are usually free from pathogens, are cheaper, and are acceptable by virtually all religious groups. On the other hand, plant proteins generally have a lower content of essential amino acids than animal protein and they are bulky, so that large quantities must be consumed to provide adequate intake. Also, they may contain toxins, and they generally lack vitamin B_{12}. Fortunately deficiencies in protein from one plant source can be balanced by mixing with that from another. Yellow pea flour is deficient in methionine, but contains adequate lysine, and the converse is true of maize meal; so a mixture provides protein of high nutritive value. Again, plant breeders can select varieties with higher contents of essential amino acids (for instance, a mutant form of maize has been discovered which has three times the normal lysine content). The problem of bulk can be overcome by extracting the protein from the plant and preparing palatable food from it. The toxins can be removed by heating or soaking the product, or developing new varieties (as with gossypol-free cotton-seed protein).

Legumes (which include the bean family) produce about three times as much protein as cereals, as well as enhancing the content of nitrogen in the soils on which they grow, by their symbiotic bacteria which can fix atmospheric nitrogen. They have been used as food by man for over 8000 years. Nevertheless, exploitation of this protein source has been far short of its potential. Selection of legumes for tropical countries, where malnutrition is severe, has scarcely begun, and much research is needed to find the varieties best suited to conditions of limited and uncertain rainfall. It is the seeds of the leguminous plants which are consumed. After a preliminary soaking, metabolic changes occur in which vitamins are synthesized by the seeds and minerals are absorbed from the water. For centuries the soya bean has been the main source of protein for Oriental peoples, its high nutritive value having been empirically recognized. Here ways have been found to overcome its disadvantages. Boiling is insufficient to make the dried beans easily edible. In China the soaked beans are emulsified, and the protein precipitated as a curd with salt and heat. The curd is much more easily digestible than the original soya bean, it is no longer bitter, and it is free from toxin (trypsin inhibitor). Fermentation of the emulsion by fungi, as in Japan, makes it more digestible and the amino acids more available. Soya bean protein, apart from methionine, is comparable in nutritive value with egg albumin and meat. There is currently considerable effort in highly industrialized countries to place the extraction of the soya bean protein on a fully economic commercial basis, and to prepare products from it which will be good to eat as well as of high nutritive value.

Large quantities of oil-bearing seeds from various types of plants have been used to meet the great demand for edible oil in the highly developed countries over the last 150 years. Ironically, it has been the countries where the inhabitants suffer most severely from protein deficiency which are the chief sources of oil seeds. Large quantities of ground nuts and cotton seeds are grown, but both are somewhat deficient in lysine and methionine. Previous methods of oil extraction wasted much of the protein in the residue which could not be used for animals or humans. The Institute of Nutrition of Central America and Panama is one of the centres looking into the urgent problem of retaining the palatability and nutritive value of the protein when removing the oil. Mixtures of protein are being marketed on an increasing scale in Latin America.

The high cellulose content of mature leaves and stems of plants makes them indigestible to human beings but there is a rapid protein synthesis in young leaves. There are now commercial operations which grind the leaves of young cereals, peas, mustard, and kale to express the sap; protein is then extracted by steam, washed, and pressed into blocks. Young leaves may well be exploited as a source of food for humans in the wet tropics.

Some seeds which could be used as a source of protein contain toxins, for instance gossypol in cottonseed, selenium in the seeds of various legumes, and a trypsin inhibitor in soya beans. The pea *Lathyrus*, which is resistant to drought and forms a large proportion of food in dry areas, contains β-amino propionitrile; this prevents the synthesis of the elastin molecule from lysine, thus interfering with the formation of connective tissue, and may cause irreversible paralysis as it affects the formation of collagen and nerve function. Some people are allergic to certain seed proteins; a group of people in the Mediterranean area, whose red blood cells are deficient in glucose-6-phosphate dehydrogenase, get jaundice if they eat broad beans.

Apart from such indigenous toxins, certain moulds associated with some leguminous seeds can be dangerous, for instance, *Aspergillus flavus* produces an aflatoxin which is fatal to poultry, even though it has not yet been proved dangerous to man.

Developing entirely new food sources

An important development in this category is the exploitation of fungi, bacteria, and algae as protein sources. The simpler forms of life have retained a synthetic capacity which higher forms have gradually lost as they evolved. From simple molecules, they are capable of building up protein which has a high nutritive value for human beings.

Yeasts are a particularly valuable source of protein. From nitrogen and inexpensive sources of carbohydrate, such as molasses or the waste from paper mills, or from the fruit, forestry, and dairy industries, yeasts yield 30 per cent of the intake as protein. Yeast is grown continuously in tanks and 75 per cent of the initial weight of culture is produced per hour. This has been carried out in the United Kingdom, Germany, and the United States. On drying the cells yield light, straw-coloured nutty or meaty flavoured flakes which are employed in sausage mixes and soup powder. A high lysine content makes them especially useful in increasing the nutritive value of proteins from plant sources.

More recently, yeast has been successfully grown on a diet consisting mainly of hydrocarbon oils with traces of nitrogen, potassium, and phosphorus. But there are difficulties; as the hydrocarbons are insoluble in water, more oxygen is required than with carbohydrates, and the reaction is exothermic, and must be cooled. However, the yield is high, 1 kg of hydrocarbon producing 1 kg of yeast. Because pure hydrocarbons are expensive to obtain on a large scale, crude petroleum, a fraction between kerosine and lubricating oil, has been employed. The usable hydrocarbon is converted to high quality protein, and the residue is a fuel oil suitable for domestic heating. It has been estimated that from 40×10^6 Mg of crude petroleum, that is 4 per cent of the current annual production, 20×10^6 Mg of protein could be produced, which is equivalent to the present annual protein production.

Up to 80 per cent of the dry weight of bacteria may be protein, making them more valuable than yeasts and other fungi; they also have a higher rate of multiplication. Recent investigations have shown that there is a strong possibility that a process can be developed whereby bacteria, supplied with a simple salt solution, can oxidize the methane of natural gas into high quality protein. The product is free from contamination with unusable hydrocarbons. Much work remains to be done, however, before it can be seen whether the process can be operated economically on a scale sufficient to make a significant contribution to world protein resources.

Algae and phytoplankton have the capacity to produce high quality protein from little more than inorganic salts, gaining their energy from light instead of from complex organic molecules. It has been calculated that 1 kg of a large fish represents the end of a biological chain produced from 100000 kg of phytoplankton. It would clearly be advantageous if man could eat phytoplankton directly, but it takes about 2.5 km of trawling to produce three tablespoonfulls of the material, which has a lobster-like flavour. There have been many attempts to cultivate algae artificially on an inorganic medium; as a result of photosynthesis *Chlorella* can synthesize a protein with taste and flavour like green tea. Under favourable conditions 20 tonnes of *Chlorella* protein could be produced in a year from an area of 4000 m^2 compared with 0.25 tonnes of soya bean protein and 0.01 tonnes of meat. At present it is not economically possible. The possibility of using algae to convert sewage into protein for feeding meat animals, and hence man, is more likely. This has been done economically, and on a fully commercial scale, in the USA.

Avoidance of food waste

A very large quantity of available food is lost because it is spoiled by microbial, insect, or animal infestation.

Seven per cent of total world crops are lost through microbial spoilage, and a further five per cent through insects. Round the Mediterranean 20 per cent of the olive crop is wasted. Chemical pesticides are being increasingly used to control infestation of growing crops, animals, and stored food commodities especially in underdeveloped areas where human survival may depend on local crop production. The principal insecticides are classified as organophosphorus (e.g. malathion), carbamates, and halogenated hydrocarbons (e.g. DDT). The principal fungicides include diphenyl, propyl, and octyl gallates, and sorbic acid. There may be adverse effects of long term ingestion of pesticides in man, and most countries, other than the United Kingdom, specify the amount of residue which is allowed. The possibility that pesticides may upset the ecological balance in plant and animal communities is receiving careful consideration. In spite of potential direct or indirect dangers to man from exposure to pesticides, their use must be considered justified where the alternative is death by starvation or malnutrition.

In areas where the diet is deficient in protein it is often as difficult to preserve food as to produce it. In such areas seasonal surpluses may be wasted because there is no means of preserving them. In many cases it is better, in the first instance, to encourage local traditional methods than to attempt scientific control. Many rural communities in underdeveloped countries have discovered that fermentation will preserve grain and fish products, presumably because of the alcohol produced. Meat is preserved by drying through exposure to the sun and air, for instance biltong in Africa and the Middle East, charqui in South America, and pemmican in the Arctic. Vacuum oil drying is a useful compromise between local custom and scientific techniques for preserving meat and similar foods. The foods are dried by immersion in heated edible oil under vacuum. The process is simple and effective.

Refrigeration could be introduced into areas where there is no fuel or water power to generate electricity. Between latitudes 40 °N and 40 °S of the equator there is sufficient solar radiation to operate refrigeration devices (for instance an intermittent absorption system).

Another way of avoiding waste is the use of essential nutrients from otherwise inedible food. Thus if spoiled fish is suitably deodorized, its valuable amino acids can be incorporated into a food which is more attractive. This has been done successfully in Central America where corn is the most important staple food; much of it is consumed as tortilla but this is deficient in lysine, methionine, and tryptophan. By incorporating three per cent deodorized fish flour with these indigenous flours INCAP has developed a nutritious and palatable food.

Fermentation can increase the nutritive value of existing foods. It increases vitamin content, makes amino acids available to the human consumer, and, as flavouring, may encourage the consumption of staple foods.

Ultimate food resources

Various amino acids have been found in investigations into the origin of life by studying the effects of electric discharges on a mixture of water vapour, hydrogen, ammonia, and methane. No doubt, means of controlling such reactions will be found. Theoretically, therefore, the nutrients required for mankind can be produced non-biologically from completely inorganic sources.

But the chemical industry has already found economical ways of synthesizing certain essential amino acids, for instance lysine, which can then be used to increase the nutritive value of lower grade foods.

Proteins have been extracted from relatively unpalatable sources and fabricated into palatable foods with added flavouring agents (for instance 'chicken', 'beef', and 'ham' which have been made from the proteins isolated from the soya bean, wheat, and casein).

It has been prophesied that by the year 2000 the synthesis of amino acids, peptides, fats, and vitamins will be economic, the production of protein and fat by micro-organisms and their extraction from green leaves and waste matter will be efficient, and desalination of sea water will be widespread enough to increase irrigation and food production.

International ethics

Scientific advances may provide adequate food for the world population in future but the social and ethical development of mankind may still prove too limited to ensure that all will eat adequately.

There are many communities where the food supplies are adequate, but not used because of long accepted traditions and beliefs. Once people have learned that undernutrition and ill health is not the normal state in adult life, there will be a desire for improvement. But food habits are deeply entrenched; whole communities must be educated to accept foods which are new or strange, and this is being attempted by the World Health Organization.

The equal right of every individual to the food necessary for life must not only be recognized, but be fully effected. It is commonsense that one-third of the world's population cannot continue to enjoy the nutrients which two-thirds of mankind lacks.

Chapter 8

Experimental investigations An introductory series

Section 1 Browning reactions in fruit

1.1 Introduction

This set of practical exercises is intended to give you the opportunity of finding out the most important changes in one deterioration of a food, the browning of fruit tissue, which is particularly noticeable in apples, pears, and peaches. It has been chosen because the changes are rapid and easily observable, and because the raw material is readily available. We shall find the conditions which affect the rate of the reaction, and then apply this knowledge to devise methods of avoiding or controlling the spoilage.

You will have observed this phenomenon when peeling and slicing apples for eating or cooking, but it is also of very great importance in large scale commercial production of fruit products. It affects the quality of apple juice, cider, apple pulp, dried fruit, canned peaches and pears, and other products.

Fruit browning reactions are relatively simple when compared with many more complicated kinds of food spoilage but the same principles apply. First find out all you can about the reaction, and then use this knowledge to devise methods of controlling the reaction.

These experiments also illustrate one of the main difficulties of food technology. This is to devise treatments which will control the unwanted reaction without producing undesirable side effects. It is no use treating apples to prevent discoloration if your treatment spoils the flavour or texture, introduces toxic material into the apples, or is so complicated and expensive that the public would not be prepared to pay for it.

All experiments with biological material should contain a control sample. If you are examining the effect of some treatment on apple tissue you should always compare the treated sample with an untreated sample of the same material. The untreated sample provides a control for the treated sample and enables you to differentiate between changes which are genuinely due to your treatment and changes which would have happened anyway without any treatment. Treated and control samples should come from the same apple, because wide variations in composition can occur between different varieties of apples, between apples of one variety at different stages of maturity, or even between apples that appear to have identical histories.

1.2 The rate of the reaction

In experiments 1.2a to 1.2e the conditions affecting the nature of the browning reaction are investigated.

Experiment 1.2a

When does the reaction occur?

This series of investigations can be conducted on one apple, provided that you know exactly what you are to do.

1 Cut a quarter from an apple, and divide the *remainder* into at least eight segments. Break one of these segments into a pulp.

Q1 Which browns faster, the pulp or the segments?

2 When the surface of the segments is brown, cut one of them in half and break another into two.

Q2 What do you notice about the freshly cut and the freshly broken surface?

Q3 What happens to these surfaces as time passes?

3 At suitable intervals cut the remaining segments in half and examine the new surfaces. Leave one segment for 24 hours.

Q4 What do you notice about the freshly cut surface as the length of time the segment has been left increases?

4 At a convenient time, bruise the quarter apple left from (1) on a hard surface, and after a few minutes cut it open and compare the damaged portion with the undamaged portion.

5 List your conclusions about the nature of the reaction.

Experiment 1.2b

Is the reaction due to micro-organisms?

Devise your own experiments, and carry out suitable ones in order to answer the question.

As a guide, if the reaction is caused by micro-organisms what could be done to kill these? Would the treatment you suggest have any effect on enzymes (in case the reaction is due to enzymes)?

Before starting any experiment check with your teacher that it is reasonable, and that the chemicals are available.

Experiment 1.2c

Does the reaction require air or some part of air?

You may well have some idea already of the answer to this question but there are a number of further tests you can make.

Design and carry out these experiments. Remember that air dissolves in water and that apples contain more than 75 per cent water.

Experiment 1.2d

What substance or substances in the apple are involved?

To answer this question from first principles we would have to analyse the apple into its various classes of compounds to see which fraction increased the darkening when added to apple pulp. This is a long and tedious process.

You will not have time to work from first principles, and so we will use some information available in the literature. Very many reactions in plant materials which lead to darkening of the tissue (for example the conversion of green tea leaves to black tea) are due to oxidation of phenolic compounds to quinones, followed by polymerization to dark brown pigments.

Apple segments could be treated with a variety of phenolic compounds to see whether any of these compounds increased the rate of browning. The following solutions must be available so that you can easily take volumes of about 1 cm^3 from them.

A 1 per cent catechol solution
B 1 per cent pyrogallol solution
C 1 per cent resorcinol solution
D 1 per cent hydroquinone solution
E 1 per cent phenol solution
F Water

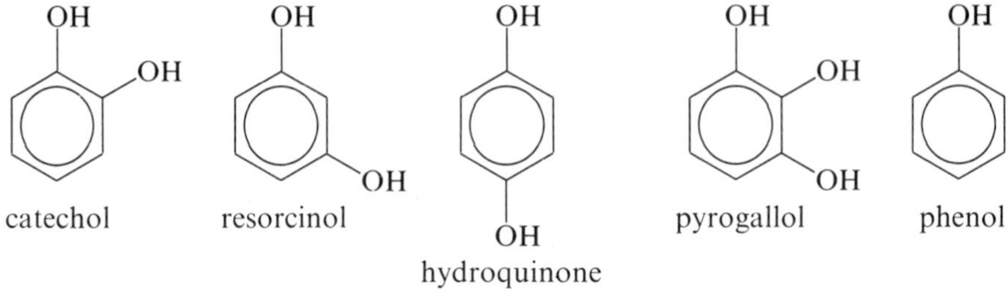

catechol resorcinol hydroquinone pyrogallol phenol

Figure 8.1

1 Cut six segments from an apple and place them separately in beakers, dishes, or on watchglasses. Wash the upper surface of the first with two portions of about 1 cm^3 of solution A; as quickly as possible wash the second segment with solution B, and so on, until the sixth has been bathed with water.

Q1 In what order is the rate of browning of the treated surfaces?

Q2 Can you find a structural feature common to the compounds whose aqueous solutions speed up browning?

2 Are the results the same if you use a pulp or juice? A liquid for testing can be made from an apple in several ways.

Either (a) Peel an apple quickly and weigh out about 50 g of undamaged material without core or skin. Homogenize with 100 cm^3 distilled water in a liquidizer or blender for 10 or 15 seconds. Strain through muslin (it is convenient if this is a bag) and use quickly.

Or (b) Prepare 50 g of apple as in (a) and as quickly as possible break it up with a pestle in a mortar containing a little clean sand. Strain through muslin and use quickly.

Add about 10 cm^3 of homogenate to each of six tubes containing respectively 1 cm^3 of solution A, 1 cm^3 of solution B, and so on. Mix well and observe at intervals.

3 Carry out the experiment as in (2) but have your solutions in 250 cm^3 beakers, or in crystallizing dishes of about 6 cm diameter.

Q3 What difference do you notice between (2) and (3)? Can you explain it?

Experiment 1.2e

Is the reaction enzymic?

Q1 What is the effect of increased temperature on the rate of a non-enzymic reaction?

Q2 What difference do you expect for an enzymic reaction?

1 Make a homogenate from an apple as in experiment 1.2d. Transfer about 10 cm^3 to each of three tubes, A, B, and C.

Tube A – immerse in a water bath at 100 °C.

Tube B – heat in a bunsen flame to boiling, and allow to boil for one minute. Cool the tube in cold water.

Tube C – keep at room temperature.

Compare the rate of browning in the three tubes.

2 Drop segments of apple into boiling water, leave them for one minute, transfer them to cold water for a minute, then leave in the air.

Q1 Is the reaction enzymic or non-enzymic?

Q2 Summarize what you know so far about the reaction.

1.3 Controlling the reaction

You should now be able to suggest methods of controlling the reaction, and further hints can be obtained by reading the beginning of chapter 4. Decide which of these ideas are practicable for solving the problem of browning on a commercial scale remembering that:

1 You may have to deal with apples in the form of juice, pulp, or segments.
2 Your product may be required to have an unaltered fresh taste or texture.
3 You must not introduce 'off' flavours, or toxic materials.

Experiment 1.3a

Can we control the reaction via the substrate?

1 Obtain samples of a number of varieties of apples including dessert and cooking apples, both English and imported. Compare the rate of browning of segments from the various samples.

2 If possible obtain apples at different stages of ripeness, and compare rates of browning of ripe and unripe apples.

Q What apples should you use to get the lightest coloured products?

Experiment 1.3b

Can ascorbic acid be used to control the browning?

Look up the structure and properties of ascorbic acid (vitamin C).
The following solutions should be ready so that you can easily and quickly take approximately 1 cm^3 samples of them.

A 5 per cent ascorbic acid solution
B 3.5 per cent ascorbic acid solution
C 2.5 per cent ascorbic acid solution
D 2 per cent ascorbic acid solution
E 1 per cent ascorbic acid solution
F 0.5 per cent ascorbic acid solution
G Water
H 2 M hydrochloric acid

1 Treat slices of apple as in experiment 1.2d, noting the time when each slice becomes noticeably browner than that treated with solution H, which acts as a standard.

2 Carry out tests on homogenate in test-tubes and/or dishes – see 1.2d (2) and (3). The exact details for these experiments will depend on the particular apples you have and the time available so discuss the matter with your teacher before you begin.

Prepare a colour standard as follows. Take 10 cm^3 of homogenate in a test-tube and allow it to stand for 10 minutes. Add 1 cm^3 of 2 M hydrochloric acid, shake to mix, and seal the tube with a bung. Mark this 'standard'.

Observe tubes or dishes containing homogenate and liquids A to H over a period of time. Note the times when the discoloration in each tube becomes:

a equal in intensity to the colour standard, and
b noticeably greater than in tube H.

Keep the tubes or dishes for 24 hours.

Q1 Does ascorbic acid stop browning?
Q2 Does it retard the reaction in proportion to the amount added?

Examine the tubes or dishes after 24 hours.

Q3 Does browning occur at a reduced rate throughout, or at its usual rate after an induction period?

Q4 From the structure and properties of ascorbic acid can you suggest how it may be acting?

3 Drop freshly cut segments of apple into a 1 per cent solution of ascorbic acid and examine after 30 minutes, 1 hour, 4 hours, 24 hours (or convenient times) by (a) noting the colour of the surface, (b) removing a segment, cutting in half, and adding a few drops of 1 per cent catechol to the freshly cut surfaces.

4 Compare the flavour of segments held under 1 per cent ascorbic acid for one hour with freshly cut segments.

Q5 Would the use of 1 per cent ascorbic acid solution be suitable for controlling the colour of apple segments?

Q6 Would it be suitable for treating segments which were being held for a few hours before being converted into pulp?

Q7 Can ascorbic acid be used to restore the fresh colour of slices which have started to turn brown?

Q8 Is ascorbic acid a suitable substance to add to food from the safety point of view?

Experiment 1.3c

Can heat treatment be used?

1 Take 5 cm^3 samples of homogenate – see 1.2d (2) – in a test tube, hold in a bunsen flame for 5, 10, 15, 30, and 60 seconds, cool the tube as rapidly as possible in cold water, and then add one drop of 1 per cent catechol. Observe whether any coloration appears.

2 Drop eight equal sized segments of apple simultaneously into boiling water. Remove a segment every 15 seconds, cool rapidly in cold water, cut in half, and add a few drops of 1 per cent catechol solution to the freshly cut surface. Draw a diagram to represent the distribution of colour after each 15 seconds.

Q1 Explain the results of these two experiments.
Q2 Can the same period of heating be used for juice, pulp, or segments?

3 Select the heat treatment you consider most suitable for controlling the colour of segments, and compare the taste and texture of segments treated in this way with unheated segments.
Q3 Is the eating quality acceptable?
Q4 Can better results be obtained at temperatures below 100 °C?
a Repeat experiment 1.3c (1), but heat the tubes in water baths maintained at 40, 60, 80, and 100 °C. At each temperature find the time required to produce no coloration with catechol.
b Take six 5 cm^3 samples of homogenate and hold in water baths at 0, 20, 40, 60, 80, and 100 °C. Measure the time taken to reach discoloration equal in intensity to a colour standard. Plot time against temperature.
Q5 Explain the shape of your graph.
Q6 Do these experiments suggest that lower temperatures might overcome the problem of 'cooked' flavour and texture? If you wish, test by further tasting experiments.

Experiment 1.3d

Can pH control be used?

Make up the following solutions:

	0.4M disodium hydrogen orthophosphate /cm^3	0.2M citric acid /cm^3	pH (approx.)
A	0.4	19.6	2.2
B	4.1	15.9	3.0
C	7.7	12.3	4.0
D	10.3	9.7	5.0
E	12.6	7.4	6.0
F	16.5	3.5	7.0
G	19.4	0.6	8.0

1 Homogenize 50 g of apple with 50 cm^3 of water – see 1.2d (2). Take 5 cm^3 portions of the homogenate and add to 5 cm^3 of each buffer solution A to G. Observe the colour formation.
Q1 Does pH affect the rate of reaction? Would you expect this result?
Q2 What is the natural pH of apple?
Q3 Should you add acid or alkali to control colour? Do you know of a 'trick' used by cooks who (unwittingly) make use of this?

2 Hold segments in (K) 1 per cent malic acid, (L) 0.5 per cent malic acid, (M) 1 per cent citric acid, and (N) 0.5 per cent citric acid.

Q4 Do these treatments provide a satisfactory solution to the discoloration problem? Why do you think these acids are suggested.

3 Check the rate of penetration of these acids into the apples.

Experiment 1.3e

The action of sodium fluoride

N.B. Add the sodium fluoride solutions from a burette or from a pipette fitted with a pipette filler. Sodium fluoride is poisonous.

1 In three test-tubes A–C place:

A 1 cm^3 of 2.5 per cent sodium fluoride solution
B 1 cm^3 of 0.25 per cent solution
C 1 cm^3 of water

then add 5 cm^3 apple homogenate – see 1.2d (2) – to each.

Q1 Does the sodium fluoride affect the rate of reaction?
Q2 Would this be a practicable way of controlling apple browning?
Q3 What do you think is the action of sodium fluoride?

Experiment 1.3f

The action of sodium hydrogen sulphite

1 In six test-tubes A–F place:

A 1 cm^3 of 6 per cent sodium hydrogen sulphite
B 1 cm^3 of 5 per cent sodium hydrogen sulphite
C 1 cm^3 of 4 per cent sodium hydrogen sulphite
D 1 cm^3 of 3 per cent sodium hydrogen sulphite
E 1 cm^3 of 2 per cent sodium hydrogen sulphite
F 1 cm^3 water

Add 5 cm^3 apple homogenate – see 1.2d (2) – to each.

Q1 Does sodium hydrogen sulphite control the rate of browning?
Q2 If so, what concentration is effective?

2 Add 5 cm^3 apple homogenate to each of two tubes A and B. Allow to stand for 15 minutes.
Then to A add 1 cm^3 of 6 per cent sodium hydrogen sulphite, and to B add 1 cm^3 of 2M hydrochloric acid.
Observe the colour at intervals for 30 minutes.

Q3 What conclusions can you draw from your results?

3 Make up two 1 dm^3 solutions of 1 per cent sodium hydrogen sulphite.

Adjust the pH of one of the solutions to pH 1.5 with 2 M hydrochloric acid. Add four apple segments to each solution. After 30 minutes, 1 hour, 4 hours, and 24 hours, remove one segment from each solution, wash in water, cut in half, and add 1 per cent catechol solution to the cut surfaces. Observe the appearance of the surfaces.

Q4 Does lowering the pH increase the rate of penetration into the apple? If so, why?

Q5 Is this a practicable method of controlling browning? Would the amount of sodium hydrogen sulphite required affect the taste?

Experiment 1.3g

The effect of solute concentration in the tissues

Certain reactions in biological tissues can be controlled by raising the concentration of solutes in the tissues. Devise and carry out a series of tests to determine whether browning in segments can be controlled by sodium chloride or sucrose without spoiling the eating quality of the apples. Explain the effects of these additives.

1.4 Summary

Summarize your findings, listing the methods which control browning. Indicate the methods which could be used commercially, and how the methods are used in a different way in home cooking.

Section 2 Cereal science

Introduction

Although some foods, such as the various fruits, can be eaten raw, most foods – meat, fish, potatoes, root crops, and cereals – are normally cooked or processed to make them palatable. For some, such as potatoes and cereals, the heating also makes digestion easier. The simplest way of cooking cereals is to boil them and that is done in many parts of the world. In Britain, however, we subject wheat for bread making to two separate processes. The first is mechanical; the grain is cleaned, ground, and sifted to separate the indigestible bran from the endosperm which is the white, particularly nutritious food store of the wheat germ. This process is called milling. In practice it is an elaborate process, since a modern flourmill has about fifty reduction and sifting steps to give a high flour yield.

During the milling of flour, very few, relatively unimportant, chemical changes take place in the flour but major chemical changes occur during storage. Some are advantageous, such as the oxidation of flour pigments by the air which gives a whiter flour. There is also a slight improvement in baking properties although it is not quite clear whether this is due to fat or protein oxidation.

This natural ageing of flour may take several weeks and the miller may add certain chemicals to speed it up. In this course you will be using some of the techniques designed to detect such additives.

Certainly, excessive fat oxidation is undesirable since it gives a rancid or 'musty' flour. This may occur if flour is stored at moisture contents significantly above 17 per cent at slightly elevated temperatures. Such deterioration is no great problem in this country, but it may be serious in tropical climates.

Baking converts flour into bread. It is a chemical change consisting of two steps. First, the flour is mixed with water and usually yeast and salt. The wheat protein, most of which is insoluble in water, takes up water and forms a three-dimensional, extensible network in which the starch grains are embedded; the yeast in the dough produces carbon dioxide as it grows and multiplies, and uses up sugars present in or produced from starch. The carbon dioxide gas is trapped by the expanding dough. Only wheat and rye give such a foam-like dough. In wheat dough the gas is held by the water-insoluble protein, 'the gluten', in rye it is held by 'mucilages'. No other cereal will give such a dough.

In the second step of the baking process the dough is converted into bread. It is a very complex and rather obscure process. The gas bubbles produced by the yeast increase in size as the dough rises in temperature. The protein network is stretched and thus the loaf rises in the oven. At about 50°C the protein undergoes denaturation, it loses water and changes from a soft putty-like state into a rubbery and tough form. The starch also loses its structure and gelatinizes. It swells by water absorption. Whereas the structure of the dough is due primarily to the protein network, the structure of the bread is due mainly to the starch. In this course you will yourself study the effect of heat on gluten protein, and observe the gelatinization of starch under the microscope. The other important changes are the browning of the crust and the production of the typical volatile materials which we associate with the aroma of a freshly baked loaf.

Experiment 2.1

Microscopic structure of wheat

The most important cereal grains are rice, which is used extensively in the Far East, and wheat which is consumed chiefly in western countries. The wheat grain is the fruit of the wheat plant. It contains cellulose, protein, starch, a little fat, and traces of valuable vitamins. As a living unit, the grain contains all the enzymes necessary for the initial stages of the life of the plant. Since wheat can utilize much simpler materials than man does, the grains (as, indeed, most plants) have far more elaborate enzyme systems than the otherwise more highly evolved animals. One of these enzymes, the presence of which is easily demonstrated, is the enzyme peroxidase which breaks down poisonous peroxides wherever they may be produced in the cell.

The object of this experiment is to find the location of cellulose, starch, protein, fat, and the enzyme peroxidase in the grain by means of histochemical tests. Histochemistry is the study of the chemical constituents of living tissues as they occur in the actual tissue.

With a razor cut thin sections of wheat grain both in the transverse and longitudinal direction. Mount in water and observe under the microscope. Draw a sketch of the principal features observed.

Having identified the anatomical features, determine their composition histochemically.

Cellulose. Place several thin sections on a slide. Add one drop of methylene blue solution to each. Leave for one minute. Mop up the dye with tissue paper. Mount in distilled water. Label your sketch with the word 'cellulose' where the material has been stained.

Fat. Place sections on a slide and wet with ethanol. Mop up as above and mount in Sudan III. Fatty tissue is stained red. Where is the fat located? Why does wholemeal flour go rancid easily?

Protein. Place a section on a slide and wet with water. Mop up the water. Add a drop of Ponceau 2 R, or other protein stain. Leave for five minutes. Flood the slide with water and let it run off to remove excess stain. Mount in water. Protein is stained red with Ponceau 2 R.

Starch. Place sections on a slide and wet with water. Leave for two minutes. Mop up the water. Mount in iodine solution. Starch is stained a dense black or blue.

Peroxidase. Mount dry sections in hydrogen peroxide and observe immediately. The enzyme liberates bubbles of oxygen. What is the chemical reaction? What would happen if you boiled the section in water for a few minutes before the test?

Experiment 2.2

Comparison of flour colour and qualitative tests for flour improvers

During the milling process the grain is broken down into flour. The endosperm is pulverized and the germ and bran are removed as flakes. White flour consists mainly of endosperm but it still contains a fair amount of bran powder. Certain vitamins are removed with the bran and germ and the miller must replace these by law. He uses a 'Mastermix' containing the vitamins together with iron in the iron(III) state. Chalk may be added in small amounts to certain flours to improve the calcium uptake in the body. The miller may also add certain bleaching agents, such as chlorine, chlorine dioxide, and benzoyl peroxide, so that the baker obtains a whiter looking loaf. Finally, 'improvers' such as potassium bromate, potassium iodate, and ascorbic acid may be added which improve the texture and volume of the loaf. Potassium iodate is not permitted in the United Kingdom at present. Ascorbic acid is permitted in almost all European countries since, as vitamin C, it is also an essential constituent of the diet.

The object of this series of experiments is to show you some differences between different flours and to determine whether the flour has been bleached or improved in any way. There are many additives and many tests for them, so only a few can be selected here. If you have some time left think how you would determine the presence of chalk in flour.

1 Comparison of flour colour

Pour flour into a Petri dish. Press down gently and, using the edges of the dish as a guide, produce a flat surface using a spatula. Compare standard bakers' flour and biscuit flour. Distinguish between background colour and bran specks.

To emphasize the difference, immerse the dishes carefully in water for one minute, sliding them smartly in at an angle of about 45°. Slide them out of the water at the same angle, drain, and dry at room temperature. Note the colour difference when still damp and when dry. To what is this colour difference due? The darkening of the wet flour is speeded up by keeping the dish warm. Why?

2 Qualitative tests for bromate and iodate

Prepare the reagent immediately before use by mixing 10 parts by volume of 2 per cent aqueous solution of potassium iodide and 2 parts by volume of 10 per cent solution of sulphuric acid. Pour the reagent over the wet flour dish as prepared above. Drain off excess reagent. What do you observe and why does it happen?

3 Qualitative test for ascorbic acid

Pour a little 2 per cent aqueous iodine solution over the wet flour in a Petri dish. Explain what you see.

In (2) and (3) you have examples of positive and negative stains. What does that mean?

4 Qualitative test for added iron

A vitamin mix is added to some flours. Since it is too laborious to determine vitamins on a routine basis, the miller adds a vitamin mix containing iron in the iron(III) state. This is then detected as follows.

Immediately before use prepare a thiocyanate reagent by mixing equal parts by volume of 10 per cent aqueous potassium thiocyanate and 2M hydrochloric acid. Drop about 1 cm^3 of the thiocyanate reagent onto the dry flour in a Petri dish and leave for ten minutes. Use untreated flour as a control. Explain what you see.

How would you modify this test if the iron were present in the iron(II) state?

Experiment 2.3

Determination of the gelatinization temperature of starch

Starch is the most important chemical component of cereals and many tubers. It is an α-glucose high polymer. The polymer chains are arranged to give spherical crystallites which constitute the starch grains. They are indigestible, and all starchy foods must be heated in the presence of water before they are eaten by man.

In the uncooked starch grain the chains of glucose molecules are very closely packed. The main bonds between them, holding the whole grain together, are hydrogen bonds. The starch grain is so stable that digestive enzymes cannot attack it. Before it can be digested by man the structure must be opened up. The most common way of doing this is to boil the starch in water. The heat breaks the hydrogen bonds, and water penetrates the grain. As a result the starch grain swells until it becomes a roundish sack filled with starch suspension. This process is referred to as gelatinization. (During baking there is only limited swelling since the amount of water is limited.)

You can observe the swelling by heating small samples of starch in water to various temperatures in test-tubes.

Mix 10 cm^3 of water and about 0.5 g of starch in a test-tube. Heat the suspension in a beaker of water at 50 °C, stirring occasionally. Cool the tube under a tap. Remove one drop of the suspension and place it on a microscope slide. Cover with a coverslip, and examine the grains for swelling and bursting. Repeat at 55, 60, 65, 70, 75, and 80 °C.

What is the gelatinization temperature of the starch which you have used?

If you have a polarizing microscope rotate the eyepiece until the background is black and each starch grain shows a Nicol cross. What is this due to? What happens to the cross on gelatinization?

During this process of gelatinization or 'jelly formation' the suspension thickens, that is, the consistency increases, and this property can also be used to determine the gelatinization temperature. In this method a steady stream of air bubbles is passed through a starch suspension which is heated. When the starch gelatinizes the air flow is suddenly restricted, and this may be observed by means of a manometer.

Starches can also be gelatinized in the cold, for example with concentrated potassium hydroxide.

Experiment 2.4

Gluten tests

Wheat grains and flour contain a large number of different proteins. When a dough is formed some of these adhere to each other and form a continuous three-dimensional network which gives coherence to the dough. Wheat proteins are the only cereal proteins to do this. These proteins are called the 'gluten complex'. When a ball of dough is washed under running water, the gluten is left behind. The first person to conduct this experiment was Beccari at the University of Bologna in 1728. At the time it was a very important experiment indeed, because it was thought that animal and plant material were basically different. Beccari showed that an animal 'element' (we now call it protein) could be obtained from vegetable material. He showed that gluten from flour was like meat, while flour was a typical plant product. His lecture 'Concerning grain' has been reprinted in the journal *Cereal Chemistry* (September 1940, page 555).

Gluten has very peculiar viscous and elastic properties which are essential for bread making. Even today after extensive research, little is known about this strange material.

Mix 56 g of one of your flour samples into a dough with 30 to 40 cm^3 of distilled water in an evaporating dish with a spatula. The dough should be homogeneous (not brittle) and have the consistency of very stiff chewing gum. Form a ball of dough and rest it for 10 minutes under water. Knead the dough ball under a slow stream of tap water (200 to 300 cm^3 per minute). The starch will be washed away as a milky suspension, and the water-insoluble 'gluten' proteins will remain. When the water is starch free (test suitably for this), squeeze out excess water from the gluten. Mould the gluten in the fingers, drying the fingers occasionally. When free water has been reduced, the gluten suddenly becomes very sticky. Place it on greaseproof paper and weigh it. Repeat for the other flour. Stretch both glutens by hand and note the consistency and springiness. Boil the gluten ball for three minutes in water. Stretch it again. What has happened chemically to the gluten? How does gluten differ from white of egg before and after boiling?

Leave uncooked gluten in water for a few days. It smells like rotting meat. This was one of Beccari's 'proofs'. How would you show that both meat and gluten contain protein?

Experiment 2.5

Baking scones

Before being consumed, cereal products must be boiled, steamed, or baked to gelatinize the starch (to make it digestible) and to coagulate (denature) the protein. The flour is mixed with sugar, fat, liquid, and self-raising ingredients. A dough is then formed and is baked in an oven. This test is used in industry to assess the wholesomeness of a flour and the effectiveness of the self-raising ingredients which are added to aerate the product. The self-raising ingredients may deteriorate during storage (how?), and the test provides a check. Most bread baked in this country is aerated by yeast but this is too time consuming for a class experiment since normally a three-hour fermentation is required.

The self-raising ingredients consist of an acid (usually acid calcium phosphate) and sodium hydrogen carbonate. This mixture will liberate carbon dioxide in the oven to raise the product. What is the chemical reaction? Pour universal indicator over a dish of *dry* self-raising flour. To what are the different coloured specks due? What is the pH of the flour itself? How would you determine total carbon dioxide in a self-raising flour? (Use dilute sulphuric acid and allow for any chalk added to the flour.)

Weigh 250 g of flour and 62.5 g of margarine into a mixing bowl and rub together. Add 62.5 g of sugar and approximately 120 cm^3 of milk to give a good dough. (The exact amount of milk depends on the flour and not too much should be used.) Mix well. Roll out the dough on a board with a rolling pin, using spacers so that the dough sheet is 1 cm thick. Cut scones with a pastry cutter and place the shapes on greaseproof paper on a baking tray. Bake for 12 to 15 minutes at 230 °C.

Repeat the experiment with the following carefully sifted into the flour, again using 250 g each time.

1 3.61 g sodium hydrogen carbonate and 3.78 g acid calcium phosphate per 300 g flour.

2 7.22 g sodium hydrogen carbonate and 3.78 g acid calcium phosphate per 300 g flour.

3 3.61 g sodium hydrogen carbonate and 7.56 g acid calcium phosphate per 300 g flour.

Measure the height and/or volume of the scones. Cut the scones and note texture and colour. Add one drop of universal indicator to a cut surface, and estimate the pH of the scone.

Also place one drop of universal indicator on some of the remaining dry flour containing the raising ingredients. Comment on what you observe.

Experiment 2.6

Boiling of dumplings

Make four different types of dumplings, using the four different recipes given below.

1 50 g plain flour only.

2 50 g plain flour, 0.6 g sodium hydrogen carbonate, 0.63 g acid calcium phosphate, mixed well.

3 50 g plain flour, 1.2 g sodium hydrogen carbonate, 0.63 g acid calcium phosphate, mixed well.

4 50 g plain flour, 0.6 g sodium hydrogen carbonate, 1.26 g acid calcium phosphate, mixed well.

Mix each of the four flours with sufficient water to give a stiff dough ball. Place the dough balls into 1 dm^3 beakers containing approximately 500 cm^3 of boiling water. Boil for 15 minutes.

Note the texture, colour, and volume of the dumplings. Place one drop of universal indicator on a cut surface and estimate the pH of the dumplings.

Chapter 9

Experimental investigations An extension series

Food research project A. A feasibility study for the large scale production of dehydrated Brussels sprouts

Your company has decided that it needs to extend its range of dried foods. Marketing research suggested that, judging by the success of frozen Brussels sprouts, there is a demand for Brussels sprouts out of season. You, as the research department, are asked to find out whether it is possible to dehydrate Brussels sprouts successfully. If this new product is to be marketed next year, decisions will have to be made very quickly to arrange for suitable raw material, and to construct pilot plant in time for next season. The Board therefore requires within one month:

1 Your opinion as to whether this new idea will work.
2 Your recommendations for the best method of processing.

After reading the chapters on food changes, you may be able to anticipate some of the problems which will arise. You will be heating a vegetable containing chlorophyll as its natural colour, and containing active enzymes at a slightly acid pH. Not only will you be able to anticipate the likely causes of deterioration but, if you have read the chapter on principles of food preservation, you will be able to suggest methods which will probably minimize the deterioration.

However, any additional stage in the process, or any additive used, will cost money and put up the price of the product. For example, if we decide to blanch the food we will have to introduce another step into the processing line, pay for fuel to heat the water or steam, and lose some of our raw material. Our initial approach should be not to anticipate these difficulties but to hope that they will not arise, and to carry out a trial run, using the simplest possible process. We can then see what changes have occurred and decide whether or not the product is acceptable. If some of the changes are not acceptable we will have to modify the process.

The final industrial process will involve the following sequence of operations.

1 Outside the factory
a Locating a suitable supply and buying the sprouts.
b Arranging for harvesting of the sprouts.
c Transport of the sprouts to the factory.

2 Inside the factory
a Storage of the sprouts until ready to process.
b Trimming the sprouts.
c Pre-drying treatment, e.g. alkaline dip, blanching.
d Drying.
e Inspection of the product and packing.
f Storage before distribution to retail outlets.

At the beginning of the development of a new product we are not able to do much about the pre-factory operations. All we can do is to ask our agricultural representative and our buyers to purchase some sprouts of reasonable quality. The only quality specifications that we can suggest at present are that the sprouts should be freshly picked, with no discolorations (external or internal), of tidy appearance, and of uniform size. In this project we need not worry any more about the pre-factory operations. However, at a later stage in development an industrial research and development unit would return to this area. Once the basic process had been established they would carry out trials to determine the optimum variety and maturity of the sprouts. They would determine the best sources of seed, agricultural conditions, and methods of harvesting, and advise the farmers accordingly. Once the desired characteristics of the raw material had been established then they might well start a plant breeding programme, designed to breed a type of sprout particularly suitable for the process.

A.1 **Tests on raw material**

Our first task is to establish a basic process, and to do this we start off with some fresh sprouts of uniform size. In the laboratory we will be able to deal with them immediately, but on a factory scale the rate of delivery from the farms would not be constant and at times it may be necessary to store the sprouts in the factory before they are processed. Our first task is to find out how long they can be stored without leading to a poorer product, and whether any particular conditions of storage are required.

Arrange with a local nurseryman or smallholder to supply a few kilogrammes of sprouts, from the same crop, each day on four successive days. Divide each day's sample into two and store one half in a warm room and the other half in a refrigerator. On the fourth day examine all the samples and compare them with the fresh sample obtained that day.

Experiment A.1a

A taste-panel assessment of the quality of the raw material

Design a tasting (and viewing) panel experiment to compare the stored samples with the fresh sample. The qualities to check are internal and external colour, odour, and flavour. The flavour is best observed in cooked sprouts. Does this need a 'difference' or 'preference' type of panel? Should the tasters know which samples they are examining? Does the order in which the samples are tasted matter? How can you quickly check the ability of the tasters? What scoring systems will you use?

When you have designed the experiment discuss its design with your teacher before you start any tasting.

After completing the experiment, you should be able to answer the following questions.

Can you detect any difference from the fresh sample? Are any different but still acceptable in quality? Are any completely unacceptable? What is your recommendation for maximum storage period?

Experiment A.1b

Chemical test for quality of raw material

We have now established maximum length of time before processing, with and without chilling. However we cannot be certain that the farmer picked the sprouts on the day he stated, and it is always possible for two different batches to be confused in the factory. We really need a test for each batch to be certain that it is fit for processing. Colour and appearance are fairly easy to judge by visual inspection, but it will be much less satisfactory, and take up too many man hours, to set up a taste panel to check the flavour of each batch entering the factory. It will be more convenient if a simple chemical test could be devised which would give results comparable to those of the taste panel.

What chemical test could we use? In chapter 3 the anaerobic breakdown of carbohydrates in vegetable tissue was discussed, particularly in vegetables damaged after harvest. Would analysis for one of these breakdown products provide the chemical test we need? Discuss the test method with your teacher.

Does the chemical test correlate with panel scores for colour and flavour?

A.2 Drying process

So far we have obtained a supply of raw materials, and determined the maximum storage time before processing. We will now start on the actual drying process. First, try drying with no preliminary treatment. Trim the sprouts, removing loose outside leaves. Discard any sprouts larger than 3.25 cm diameter. Drying times will probably vary with size and so it is best to divide the sample into three grades, small, medium, and large. Lay out the sprouts in thin layers on a tray and place them in an oven at 65 to 70 °C for $2\frac{1}{2}$ hours; then reduce the temperature to 50 °C and continue drying for 12 hours. If an oven with a circulating fan is available it will give better results. Check that the moisture content is 6 per cent or less. If the moisture content is more than 6 per cent continue to heat at 50 °C for another two to three hours.

Experiment A.2a

Determination of moisture content

Mince or grind the product, weigh out a 5 g sample into weighed beaker, sufficiently large to allow the product to be in a thin layer, and dry at 100 °C in an oven (air-circulation oven if available) for $1\frac{1}{4}$ hours. Cool in a dessicator and weigh again when cool. Make sure that the sample is representative of the centre as well as of the outside of the sprouts.

Experiment A.2b

Examination of dry product

Rehydrate the sprouts. Add 10 g of dried sprouts and 4 g of sodium chloride to 500 cm^3 of cold water, bring to the boil, and simmer for six minutes. Cut the sprouts in half and examine the interiors as well as the exteriors.

Examine a dehydrated sprout cut in half. Has it dried in the centre? Check this by comparing the moisture content of the outer leaves with the centre. Is the condition of the centre related to the size of the sprouts? Do you think the appearance of the product (after rehydration) is satisfactory? Is the colour satisfactory?

If the colour is unsatisfactory, what are possible causes? After reading chapter 3 you should be able to list three possible causes. What method can you suggest to decide which of these is responsible? Suggest ways of treating the sprouts so as to improve the colour. Discuss your suggestion with your teacher before proceeding.

A.3 Blanching

Nearly all industrial methods of preserving vegetables (whether canning, freezing, or drying) include a blanching process. What is the purpose of blanching?

Experiment A.3a

Blanching method

Boil 400 cm^3 of water in a saucepan or 1 dm^3 beaker, tip in 20 g of sprouts, and hold at 98 °C for the required time. Strain off the sprouts and tip them into 400 cm^3 of cold water. Strain off the sprouts again. A batch of sprouts should be graded into three size grades, and samples from each grade blanched for 1, 2, 3, and 5 minutes. Determine whether each sample has been adequately blanched by testing for the enzyme peroxidase.

Experiment A.3b

Peroxidase test

Mash up about 10 g of blanched sprouts. Take a 5 g sample and grind with sand and 5 cm^3 distilled water in a mortar. Decant the mixture into a 2.5 cm diameter test-tube. Add 1 cm^3 of 0.5 per cent hydrogen peroxide and 1 cm^3 of 1 per cent alcoholic guaiacol solution and allow to stand for five minutes. If a red-brown coloration has appeared by the end of five minutes, active peroxidase is still present in the sample.

Is there any relationship between the size of the sprouts and the time needed to give a negative peroxidase test?

Take one sprout from a sample which still shows some peroxidase activity. Cut it in half and pour a few drops of 1 per cent guaiacol onto the cut surface followed by a few drops of 0.5 per cent hydrogen peroxide.

What does this tell you about the time needed to inactivate all the peroxidase?

The optimum size of sprouts for dehydration

You will probably now have met three difficulties with the larger sprouts: drying, rehydrating, and blanching the centres. How do we overcome these problems? One apparently simple solution would be to use only those sprouts under a certain size. However this would mean throwing away the larger sprouts, or re-selling them as low grade sprouts, either of which would put up the cost of the final product. In another year we could possibly arrange for the sprouts to be cut at an earlier stage in their growth, or to breed a variety in which the sprouts did not grow beyond a certain size. Another possible solution is to cut the large sprouts in half. We can foresee two snags. Firstly, the customers may expect whole sprouts and reject half-sprouts. Secondly, a lot of solids and flavour may be lost if a half-sprout is blanched. There is nothing we can do about the first at present but we can examine the blanching losses.

Experiment A.3c

Determination of blanching losses

1 Solids content of unblanched sprouts

Take 50 g of sprouts and chop or mince them into fine pieces. Weigh out accurately about 40 g into a beaker and dry at 90 °C overnight in an oven. Weigh them again, grind up the sample, and weigh 5 g into a beaker; dry at

100 °C for $1\frac{1}{4}$ hours. Weigh again. Calculate the solids content of the original sprouts.

2 Solids content of blanched sprouts

Weigh out 50 g of sprouts and blanch in 1 dm^3 of water for three minutes at 98 °C. Strain off the sprouts and cool in 1 dm^3 of cold water. Strain off the sprouts and allow the surface water to drip away. Roll the sprouts gently on blotting paper to remove excess water. Chop up the sprouts and determine the solids content as in (1). Repeat with 50 g of halved sprouts.

From your results you can calculate the following quantities:
Weight of dried product (at 6 per cent moisture content) obtainable from 100 g of sprouts without blanching = x g.

Weight of dried product (at 6 per cent moisture content) obtainable from 100 g of sprouts if they are blanched whole = y g.
Weight of dried product (at 6 per cent moisture content) obtainable from 100 g of sprouts if they are blanched after halving = z g.

Then loss of product due to blanching

$$= \frac{x-y}{x} \times 100 \text{ per cent.}$$

Additional loss of product due to halving

$$= \frac{y-z}{x} \times 100 \text{ per cent.}$$

These figures will have to be reported at the end of the work so that the company's financial department can assess whether it is more economical to bear this processing loss, or to re-sell the large sprouts.

3 Flavour of halved sprouts

Carry out a taste panel experiment to compare the flavour of:

a Dehydrated unblanched whole sprouts.
b Dehydrated blanched whole sprouts.
c Dehydrated blanched halved sprouts.

Make recommendations, based on these results, for the maximum size of sprouts which can be used, and the optimum blanching time. Will halved sprouts give a satisfactory product?

A.4 Treatment with additives

Alkalis and sulphite are the most common additives used in preserving vegetables. What is the function of these additives? Design and carry out experiments to determine whether they can be used to improve the quality of dehydrated sprouts. Some hints on experimental details are given below.

Alkali. Sodium carbonate can easily be used to raise the pH. The best point at which to add ingredients to vegetables is normally immediately after blanching and the alkali dip can be combined with the cooling stage.

Sulphite. Sodium sulphite is the most convenient to use. This should be added to the alkali dip. The final concentration is fixed by three factors: firstly, there must be sufficient sulphur dioxide in the product to prevent browning; secondly, the legal maximum of sulphur dioxide in the dehydrated vegetables in this country is 2000 parts per million; thirdly, if too much sulphur dioxide is present it will be detectable as an 'off' flavour.

Experiment A.4a

Determination of the optimum concentration of sulphite and carbonate in the dip

Blanch 200 g of sprouts (small or halved) in 4 dm^3 of water, strain, and tip them into 4 dm^3 of cold water and stir gently for two minutes; then strain and dry them in an oven as before. Add varying quantities of sodium sulphite (from 0.3 to 1.2 per cent) and sodium carbonate (0.2 to 1.0 per cent) to the cooling water. A panel should examine each batch for colour and flavour. Analyse them for sulphur dioxide, and determine the pH.

The sulphur dioxide content should be determined on the dried sprouts before rehydration. The pH of the water in which the sprouts are rehydrated will give a sufficiently accurate measurement of pH. A pH meter and glass electrode should be used, but, if these are not available, use indicator papers.

Experiment A.4b

Determination of sulphite as sulphur dioxide

Reagents

Orthophosphoric acid, 85 per cent solution
Bromophenol blue indicator, 0.1 per cent *w/v* in 50 per cent *v/v* ethanol
Hydrogen peroxide, 3 per cent solution. Add 10 cm^3 of bromophenol blue indicator to 50 cm^3 of 30 per cent hydrogen peroxide ('100 volume') and neutralize with barium hydroxide solution. Allow to stand overnight and filter. Dilute to 500 cm^3 with distilled water.
Sodium hydroxide, 0.1M, standardized using bromophenol blue indicator

Set up apparatus for steam distillation, using a 1 dm^3 flask as the distillation flask and a 3 dm^3 flask for the steam supply. The condenser should be fitted with an adaptor reaching almost to the bottom of the collecting flask (500 cm^3) so that the tip is covered when there is 25 cm^3 of liquid in the flask.

Grind up about 30 g of dried sprouts. Weigh out 20 g accurately.

Measure 300 cm^3 of distilled water and 20 cm^3 of orthophosphoric acid into the 1 dm^3 flask, mix well, and introduce the weighed sample. Immediately pass steam through the mixture and apply heat to the flask until boiling has started. Collect the distillate in 25 cm^3 of 3 per cent hydrogen peroxide (adaptor below surface). After 15 to 20 minutes, when about 200 cm^3 distillate has been collected, stop the distillation, wash down the end of the adaptor, and titrate the distillate with 0.1M sodium hydroxide solution. A blank determination should be done for each batch of reagents.

You should now have sufficient information to suggest a process which will give the optimum results.

A.5 **Nutritive value**

Finally we need to know whether the nutritive value of the sprouts has been lowered too severely by the process.

Which components should you analyse in order to measure changes in nutritive value of the processed sprouts? Remember that sprouts are mainly carbohydrate, but are also a valuable source of one vitamin. What carbohydrates are most likely to be lost in the process?

What standard should you use when assessing the nutritive value of the processed sprouts? Discuss the design with your teacher, who will give you details of the analysis to be used.

A.6 **The final report**

You should now be able to write your report, giving the most promising set of conditions to be tried out on a pilot plant, and the recommendations for any further laboratory work you consider desirable. The product would of course require storage trials before any decision could be made on pilot plant experiments, but you will have no time in which to carry out these trials.

If you feel the product is so unpromising that the project should proceed no further, give your reasons in your report.

Food research project B. An improvement study for the large scale production of frozen potato chips

Your company has decided to enter the frozen food industry and has started by producing a limited range of products. They wish to market frozen chips, but pilot scale studies have revealed hitherto unsuspected problems in the pre-freezing stages of processing and given a product considerably inferior to those of leading competitors in the field.

You are asked to examine your own company's product and that of the competitor, up to, but not including, the freezing stage. You are asked to specify the differences, find the reason for them, and to suggest modifications of the process in order to give a better selling product.

B.1 Initial examination of the chips

You are given the following information by your own company.

The chips are produced from potato tubers (variety Majestic) which have been stored at between 2 and 5 °C, for at least four weeks. The chips are prepared in a chipping machine (hand or mechanical) which cuts along the longitudinal axis of the potato. They are fried in hardened cottonseed oil at 190 °C not exceeding a 1:6 ratio of chips to fat. Heating is continued at that temperature until the chips rise to the top and then for a further minute (about $3\frac{1}{2}$ minutes in all). For this particular batch of chips, old, that is well used, frying oil should be used. Cool the chips to room temperature and then reheat using the technique recommended for the commercial product you are using as a standard.

For the purposes of comparison, the competitive product (which should be a reliable brand) should be reheated in fresh cottonseed oil according to the recommendations on the package.

Examination

1 Examine the texture of the product. The texture of the chip may be objectively measured by noting the angle subtended on a protractor by the extreme free end of an 8 cm chip held horizontally at one end in a clamp (see figure 9.1). This should be repeated with a total of six chips from the same batch. Note any internal difference in texture when a chip is broken open.

2 Note any difference in colour between your chips and those of your competitor.

3 Heat the chips in fresh cooking oil and compare their tastes.

Tabulate your results for the two products under the headings of texture, colour, and taste, to show why your product is inferior.

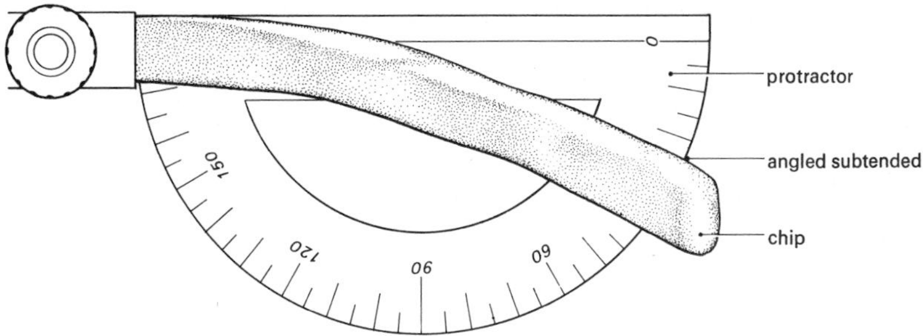

Figure 9.1

B.2 **Texture**

The texture of the chip is related to its cellular structure, particularly the condition of the cell walls and the pectin of the middle lamella. If de-esterification (i.e. hydrolysis of the methyl groups) of the pectins takes place in the presence of positive ions (particularly calcium ions, either present naturally or added during processing), interaction between these and the free carboxyl groups occurs and there is an evident increase in the rigidity of the cell walls. Such de-esterification is catalysed by the relatively heat-stable pectin methyl esterase which is present in potatoes. In fact, heating to 55 to 60 °C results in a considerable enhancement of activity.

A further important factor is the effect of elevated temperatures upon the intracellular starch granules while in an aqueous environment. At temperatures of 55 to 60 °C, the starch present in the potato cells begins to swell through absorbing water, which also increases the internal pressure and tends to round off the individual cells by breaking the intracellular cement without rupturing the cell wall. This rounding off of the cell and the associated strengthening of the cell wall by heat treatment leads to a floury texture which, though not necessary in potato chips, is essential in some products. Individual potato varieties differ in their contents of pectin methyl esterase and starch, and indeed in the inherent strength of the cell walls.

The aim therefore is to choose a potato variety which, under suitable conditions, will give chips of the desired overall texture. The following experiments will show how variation of the pre-cooking procedure will result in a different type of product, one of which may compare quite closely with that of your competitor.

Use King Edward and Majestic potato varieties in the following experiments.

Experiment B.2a

What is the effect of various preheating regimes on the two varieties, before blanching all samples for one minute at 95 to 100°C?

You may find it more convenient to use chips of circular cross-section produced with a cork borer in the following treatments.

1. Preheat for 5 minutes at 55 °C; then blanch for 1 minute at 95 to 100 °C.
2. Preheat for 10 minutes at 55 °C; then blanch for 1 minute at 95 to 100 °C.
3. Preheat for 15 minutes at 55 °C; then blanch for 1 minute at 95 to 100 °C.
4. Preheat for 5 minutes at 75 °C; then blanch for 1 minute at 95 to 100 °C.

Repeat (1) using 0.25 per cent solution of calcium chloride for the preheating regime.

Fry the samples under the usual conditions and examine after ten minutes. Tabulate your results.

Experiment B.2b

What is the relative pectin methyl esterase content of the two potato varieties?

1 Extract the enzyme

Blend peeled potatoes (400 g) for one minute with 400 cm^3 of 2.0 M sodium chloride, and adjust the pH (from about 5.7) to 8.0 with sodium hydroxide. Allow the blend to stand at room temperature for three to six hours to de-methylate the potato pectin, maintaining the pH at 8.0 by addition of alkali at intervals. Leave overnight at 0 °C. (This improves the extraction.) The extract darkens and goes black within a few hours. The pectin methyl esterase is separated from the pulp by centrifugation (or straining through butter muslin), and stored under toluene at 0 °C.

2 Determination of pectin methyl esterase activity

The substrate is a solution of commercial Pectin N.F. (0.5 per cent pectin + 1.03 per cent sodium chloride in distilled water). The enzyme is the crude aqueous sodium chloride extract made in (1).

Mix solutions of pectin (100 cm^3), sodium acetate (5 cm^3), sodium oxalate (5 cm^3), and sodium chloride (10 cm^3), in a 250 cm^3 beaker. Place the beaker in a constant temperature bath at 30 °C and stir continuously (preferably with a magnetic stirrer). Using a pH meter and glass electrode, adjust the pH to approximately 7.0 with sodium hydroxide. Add the enzyme solution (10 cm^3), adjust the pH to exactly 7.0, and start a stopwatch. Keep the pH at 7.0 by adding 0.05M sodium hydroxide and record the alkali used at five minute intervals. The run is continued for 30 minutes.

A graph may be plotted of cm^3 alkali added against time.

The relative pectin methyl esterase activities of the two samples are given by the gradients of the graphs of the two extracts. (This can also be expressed by comparing the number of cm^3 of 0.05 M sodium hydroxide required to maintain the preparation at pH 7.0 for a specific time.)

Do calcium ions have any noticeable effect upon texture?

Can the relative pectin methyl esterase content of the two varieties of potato be related to their susceptibility to induction of textural changes through preheating?

B.3 **Darkening of chips**

Darkening of vegetable material may arise from a variety of sources, two of the most important being enzymic browning, and non-enzymic browning.

Enzymic browning. A typical example is the following.

tyrosine ($CH_2—CH(NH_2)—CO_2H$, OH) —a polyphenol oxidase→ 3,4-dihydroxy phenylalanine (DOPA) ($CH_2—CH(NH_2)—CO_2H$, HO, OH)

↓

DOPA-quinone ($CH_2—CH(NH_2)—CO_2H$, O, O) —polymerization→ $[\ldots]_n$

Non-enzymic browning. This is a very complex change. The Maillard reaction is an important component of the colour change. This is described on page 56.

This is a reaction which is involved in the surface darkening of potato chips and may be arrested by the presence of sulphur dioxide or sulphites. The darkening of the internal flesh of the chips proceeds by a different mechanism altogether. This, however, is also inhibited by the presence of sulphur dioxide or sulphites and by high concentrations of citric acid.

Both enzymic and non-enzymic browning of the chip surfaces may occur in the potato. It is desirable therefore to determine whether both or only one is contributing to the problem in the chips.

Experiment B.3a

Are enzymic processes responsible for the discoloration of the potato chips?

The enzyme responsible for darkening in potatoes is polyphenoloxidase, and its effect can be seen on the surface of uncooked potatoes exposed to air. We must determine whether the preheating and cooking procedures recommended result in the destruction of polyphenoloxidase.

Determine the polyphenoloxidase activity in Majestic potatoes after the following treatments:

1 Untreated.
2 Preheated for 5 minutes at 55 °C.
3 Preheated for 5 minutes at 55 °C; blanched for 1 minute at 95 to 100 °C.
4 As for (3), but cooked at 190 °C as in the chip frying technique.

Extract 500 g of potato into 1 dm^3 of phosphate buffer at pH 7.0 using the blender. (The phosphate buffer is conveniently prepared by mixing 390 cm^3 of 0.2 M NaH_2PO_4 with 610 cm^3 of 0.2 M Na_2HPO_4 and making up to 2 dm^3.) Strain through muslin. Take a 0.1 cm^3 portion and add to 5 cm^3 of 0.001M 3,4-dihydroxyphenylalanine in phosphate buffer. Plot the rate of colour development with time, using any simple colorimeter (choose the filter giving maximum absorption). Calculate activity from the slope of the graph, which should be linear over the first five minutes, using the untreated potato (1) as the standard.

Do you foresee enzymic darkening being a problem in commercial production of chips?

Experiment B.3b

Is non-enzymic darkening responsible for the discoloration of the potato chips?

1 What are the relative effects of different sugars on non-enzymic darkening of chips?

Soak blanched potato chips for one hour in a 1 per cent solution of glucose, fructose, and sucrose, and then fry them. Note any differences in colour.

2 Does the temperature of storage affect the sugar content of potatoes?

Determine the sugar content of potatoes stored for at least one month at 2 to 5 °C and those stored in a cool room by the technique outlined below. The use of paper chromatography, time permitting, will allow a rough separation and analysis of the sugars to determine their relative concentrations.

Extraction

Blend 200 g of potatoes adding ethanol gradually during the blending until the final ethanol concentration is 80 per cent (taking into account the fact that potatoes are approximately 80 per cent water). If no blender is available, grind the potatoes with sand in a mortar under the solvent. Remove the insoluble material by centrifugation (or by straining through two layers of butter muslin) and wash several times with 80 per cent ethanol. Combine and evaporate the supernatant and washing on a water bath preferably at reduced pressure. Partition the resulting syrup in chloroform:methanol:water (8:4:3). The upper methanol:water layer contains all polar substances such as sugars, amino acids, and organic acids, and the lower chloroform:methanol layer contains lipid material. Evaporate the upper fraction to dryness.

Take up the residue in 50 per cent aqueous alcohol to give a concentration of about 160 mg cm^{-3}. Portions of this solution can be used for the sugar analyses.

Determination of sugars

a Total sugar

The phenol/sulphuric method can be recommended for sugar estimation since it is sensitive, rapid, and accurate. It estimates all sugars whether reducing, non-reducing, substituted, or in polymeric form. In potatoes, after storage, sucrose, fructose, and glucose, together constitute the majority of the sugars.

Put 2 cm^3 of a sugar solution containing between 0.01 and 0.1 mg cm^{-3} of sugar into a test-tube or colorimeter tube and add 1 cm^3 of an aqueous solution of phenol. Add 5 cm^3 of concentrated sulphuric acid rapidly, directing the stream of acid against the solution surface in order to obtain good mixing. Let the tubes stand for ten minutes, then shake them and place them in a water bath for 10 to 20 minutes at 25 to 30 °C before taking readings. A colour filter should be chosen which transmits the wavelength that is most strongly absorbed by the solution. This will be around 4.80×10^{-7} m. The colour is stable for several hours and readings may be made later if necessary.

A standard curve may be prepared for solutions containing 0.02, 0.04, 0.06, 0.08, and 0.1 mg cm^{-3} sugar. If no colorimeter is available then the solutions of unknown concentration may be compared visually with those of known concentrations.

For determination of the total sugars, a standard curve for the phenol/sulphuric acid reaction may be prepared from a mixture containing equal proportions of glucose, fructose, and sucrose over a total concentration range of 0.01 to 0.1 mg cm^{-3}.

b Quantitative estimation of glucose, fructose, and sucrose

A portion (0.25 cm^3) of the sugar solution to be assayed is streaked onto a paper chromatogram (Whatman No. 1 is suitable) and the chromatogram developed (either ascending or descending) for about six hours in ethyl acetate:acetic acid:water (10:5:2). Spots of the individual sugars (as 10 per cent solutions in 50 per cent ethanol/water) are also applied and run as markers for the bands. Strips from the centre and sides are removed and sprayed with p-anisidine and heated to 100 °C for five minutes (this time should not be exceeded).

The bands of sucrose, glucose, and fructose thus located on the untreated chromatogram are cut and the sugars eluted with water. The sugar in each sample is estimated by the phenol/sulphuric acid method.

3 Does preheating or blanching affect the surface darkening of fried chips?

Use potatoes of Majestic variety which have been stored at 2 to 5 °C to compare the surface browning of those that have been cooked without any pre-treatment or blanching with those which have received the following different treatments of preheating and blanching.

a Preheating for 5 minutes at 55 °C and blanching for 1 minute at 95 °C.

b Preheating for 15 minutes at 55 °C and blanching for 1 minute at 95 °C.

c Preheating for 5 minutes at 55 °C in 0.1 per cent sodium sulphite and blanching at 95 to 100 °C for 1 minute also in 0.1 per cent sodium sulphite.

Tabulate your results.

Discussion

Do you think low temperature storage will lead to non-enzymic browning of the potatoes?

Is there any suggestion that the concentration of one (or more) sugar in potatoes is likely to be critical in the development of non-enzymic browning?

How do you think preheating and blanching exercise a beneficial effect, if they do?

Does sodium sulphite exercise its effect in the same way?

B.4 Rancid and 'off' flavours of chips

The frying of chips at a high temperature (190 °C) and the release of water vapour from them results in deterioration of the frying oil. Chips fried in old oil are usually poorly flavoured and have a relatively short shelf life.

Water in the frying oil results in hydrolysis of glycerides to diglycerides and free fatty acids, and the presence of air leads to oxidation and rancidity. Diglycerides, monoglycerides, and free fatty acids are formed:

$$\begin{array}{l} CH_2O_2CR^1 \\ | \\ CHO_2CR^2 \\ | \\ CH_2O_2CR^3 \end{array} \xrightarrow{H_2O} \begin{array}{l} CH_2OH + R^1CO_2H \\ | \\ CHO_2CR^2 \\ | \\ CH_2O_2CR^3 \end{array} \xrightarrow[\text{(two stages)}]{H_2O \quad H_2O} \begin{array}{l} CH_2OH \\ | \\ CHOH \\ | \\ CH_2OH \end{array} + \begin{array}{l} R^1CO_2H \\ R^2CO_2H \\ R^3CO_2H \end{array}$$

triglyceride — diglyceride — glycerol — free fatty acids

Peroxide formation takes place most easily in unsaturated glycerides (hence the use of hardened, that is, hydrogenated, vegetable oils to prevent its occurrence), and is an intermediate stage in the formation of undesirable carbonyl compounds which, however, are more difficult to determine quantitatively. Hence the peroxide value may be used as a helpful indicator of recent exposure to oxidizing conditions but is not a critical criterion of oxidative rancidity.

$$RCH{=}CHCH_2R^1 \xrightarrow{O_2} \begin{array}{l} O_2H \\ | \\ RCH{-}CH{=}CHR^1 \end{array} \longrightarrow \text{carbonyl compounds responsible for rancidity}$$

unsaturated triglyceride — organic peroxide intermediate

Experiment B.4a

What are the free fatty acid contents of old frying oil and a fresh sample of hardened cottonseed oil?

The acid value of an oil or fat is the number of milligrammes of potassium hydroxide required to neutralize the free fatty acids in one gramme of the material. The acidity may also be expressed in terms of the percentage of free fatty acids present, calculated as the free acid predominating (oleic acid, palmitic acid, or lauric acid).

Reagents

Ethanol 95 per cent, 0.1M or 0.5M aqueous sodium hydroxide, 1 per cent phenolphthalein solution in ethanol.

Procedure

Weigh a suitable quantity of the sample into a 250 cm^3 flask, the quantity being sufficient to give a final titration of about 10 cm^3, and usually between 2 and 15 grammes. Using a piece of porous pot, bring 50 cm^3 of ethanol to the boil; add 1 cm^3 of the phenolphthalein solution, and titrate to a faint pink while still hot. Add the neutralized ethanol to the 250 cm^3 flask, mix the contents, and bring them to the boil. Titrate, while still hot, with 0.1M or 0.5M sodium hydroxide solution until a faint pink colour persists for at least ten seconds. Shake the contents of the flask continuously and as violently as possible during the titration (a glove with which to hold the flask may be useful or a magnetic stirrer incorporating a heater), since in general the oil is not soluble in the alcohol and the free fatty acids have to be brought from the oil phase to the alcoholic one. The titration should, as mentioned, be of the order of 10 cm^3. Adjust the strength of alkali and weight of oil used accordingly.

Experiment B.4b

What are the oxidative rancidities of old and fresh cooking oils?

The peroxide values give a useful indication of oxidative rancidity. The presence of peroxide oxygen in an oil results from autoxidation, and incipient rancidity is usually detected by an iodometric method. There are various methods available of which the following is one.

Dissolve 5 ± 0.05 g of the oil in 30 cm^3 of a mixture of glacial acetic acid and chloroform (3:2 by volume) and add 1 cm^3 of saturated potassium iodide solution. Stir the mixture by giving a rotary motion to the flask. Exactly one minute after the addition of the potassium iodide, add 100 cm^3 of water and titrate the liberated iodine with 0.1M or 0.01M sodium thiosulphate solution, depending on the amount of iodine liberated. The end-point is obtained by the use of starch as an indicator. Vigorous shaking is necessary to remove the last traces of iodine from the layer of chloroform.

Calculate the percentage of free fatty acids present in the oil. It is customary to record the percentage of free fatty acids present in palm oil as palmitic acid (molecular weight 256), in coconut and palm kernel oils as lauric acid (molecular weight 200), and in other oils as oleic acid (molecular weight 282). The acidity of oils of the rapeseed group is often expressed in terms of erucic acid (molecular weight 338) in Europe.

Calculate the peroxide values of the two oils in terms of cm^3 of 0.002M thiosulphate per gramme of oil.

Discussion of results

Is there any correlation between the flavour of the chips and the free fatty acid content and peroxide values of the oils?

Why is the peroxide value not necessarily a valid indication of the degree of oxidative rancidity?

B.5 **Heat penetration studies**

It may also be of interest to follow the temperature changes at the centre of a chip during the various preheating, blanching, and frying operations, in particular in the experiments connected with changes in texture. This may be conveniently accomplished through the use of thermocouples. It is important to ensure that the thermocouple is inserted into the chip so as to give maximum insulation, otherwise the effect of heat conduction along the wire will override the effect of conduction through the tissue.

B.6 **Conclusion**

From your results prepare written recommendations for the manufacture of chips of the potato variety Majestic. Make recommendations also on the suitability of King Edward potatoes for chip production and suggest any modifications that might be necessary in the production schedule.

Appendix
Further reading for chapter 6, Food legislation and public health

The first three books are easy to understand and provide a good general background.

Drummond, J. C. and Wilbraham, A. (1958) *The Englishman's food*, Jonathan Cape, London.
Amos, A. J. (Ed.) (1960) *Pure food and pure food legislation, 1860–1960,* Butterworth, London.
Royal Society of Health (1962) *Food poisoning*, a symposium, London.

The following official papers are of interest for reference. A selection of reports and regulations is given.

Reports of the Food Standards Committee, HMSO:
Bread and flour (1960)
Flavouring agents in food (1965)
Claims and misleading descriptions (1966)
Regulations, HMSO:
The soft drinks regulations (1964)
The bread and flour regulations (1963)
The arsenic in food regulations (1959)

Other sources

The Food and Drugs Act (1955) HMSO
Herschdoerfer, S. M. (Ed.) (1967) *Quality control in the food industry*, **1**, Academic Press, London and New York. (A general reference book.)
O'Keefe, J. A. (1956) *Bell's sale of food and drugs*, Butterworth, London. (Standard reference for regulations and legal cases.)

Index

References to specific salts are indexed under the name of the appropriate cation; references to substituted organic compounds are indexed under the name of the parent compound.